Understanding Special Relativity and Maxwell's Equations

With Implications for a Unified Field Theory

Richard E. Haskell

ISBN: 978-1516864744

Published by Richard E. Haskell, Inc.
Printed by CreateSpace, An Amazon.com Company

Preface

The order in which scientific laws are discovered is not predetermined. It isn't the case that one discovery inevitably leads to the next in a logical order. Occasionally, a disruptive theory comes along – the special theory of relativity, quantum mechanics – that radically changes the way scientists think about the laws of physics. New theories, based on these discoveries, are then followed down paths which may or may not lead to a unification of all physical theories. Perhaps a wrong turn was taken some time ago, which is the cause of the inability, so far, to reconcile quantum mechanics with general relativity, for example, and create a unified field theory that explains everything. And what ever happened to classical electromagnetic theory? We know it is central to almost everything in our everyday lives – radio, TV, cell phones, and, in fact, everything electrical or magnetic. Doesn't it make sense that classical electromagnetic theory should play a central role in any unified field theory? Instead, most of the effort has concentrated on combining general relativity, which is concerned with gravity – a much weaker force than electromagnetic forces – and quantum mechanics, of which many of its greatest contributors, including Albert Einstein and Erwin Schrödinger, were uncomfortable with its strange connection to physical reality.

As an example of the order of scientific discoveries, consider Maxwell's discovery of the laws of electrodynamics in the 1860s. He discovered his equations by studying the experiments of electrical phenomena carried out by Michael Faraday during the 1830s and 1840s. Maxwell's equations predict that electromagnetic waves travel in free space with a velocity equal to the speed of light. Maxwell therefore believed that light was a form of electromagnetic waves, but most scientists at the time were skeptical. No one knew how to generate or detect electromagnetic waves. Maxwell died in 1879 at the age of 48, eight years before Heinrich Hertz generated and detected electromagnetic waves for the first time in his laboratory at the Institute of Technology in Karlsruhe, Germany. In 1905, Einstein published his special theory of relativity.

Suppose that Einstein was born 100 years earlier. Could he have developed his special theory of relativity in 1805 instead of 1905? While it is true that Einstein's 1905 paper was called *On the Electrodynamics of Moving Bodies*, and referred to Maxwell's electrodynamics in the first sentence, the theory itself rests on only two assumptions: The first, which he calls the *Principle of Relativity*, is that there is no such thing as absolute rest and that absolute uniform motion cannot be detected by any experiment. The second assumption is that the speed of light is independent of the motion of the light source. The speed of light had been estimated by James Bradley in 1728 using stellar aberration. His estimated value was within about a half of one percent of the current value of the speed of light. Inasmuch as Einstein's 1905 paper contained no references, and was based to a large extent on mental experiments, it is conceivable that Einstein might have come up with it in 1805. At that time, Einstein would have known of Charles Coulomb's 1784 experiment showing the inverse square law for the force between two charged balls.

In this book, we will show how Maxwell's equations of electrodynamics can be derived from only Coulomb's law and Einstein's special theory of relativity. This means that if Einstein lived 100 years earlier, perhaps someone could have derived Maxwell's equations without reference to any of Faraday's experiments. Who might that person be? As we will see, the derivation requires some vector analysis, but today's form of standard vector analysis was not introduced until the 1880s by Josiah Willard Gibbs and Oliver Heaviside. However, their vector analysis was really a subset of the more general quaternions discovered by Sir William Rowan Hamilton in 1843. Hamilton, who was born in Dublin, Ireland in 1805, was arguably one of the greatest mathematicians of his time. Hamilton thought that quaternions were the best type of mathematics to describe all physical theories, and spent the last 22 years of his life studying quaternions. So if Einstein lived 100 years earlier, perhaps Hamilton would have used his quaternions to derive Maxwell's equations. In fact, using quaternions, Maxwell's normal four equations reduce to a single equation. We will show how this is done in Section 4.7.

I first came across the derivation of Maxwell's equations from Coulomb's law and special relativity in 1966 in an article by R. S.

Elliott in the IEEE Spectrum. Elliott also included the derivation in his *Electromagnetics* book that same year. I was just completing three years in the Air Force working in the Microwave Physics Laboratory at the Air Force Cambridge Research Laboratories in Bedford, MA. A colleague of mine, Carl T. Case, and I were busy solving Maxwell's equations for electromagnetic waves propagating through plasmas (ionized gases) in a magnetic field when we came across the derivation. We redid the derivation using our own notation – which is the derivation given in this book. In the latter part of 1966, I joined the newly formed engineering school at Oakland University in Rochester, Michigan.

I started teaching courses in electromagnetic theory at Oakland University and I would include this derivation of Maxwell's equations in these classes. Carl and I realized that the derivation assumed the assumption of special relativity that the moving reference frame was moving with a constant velocity. Thus, we were pretty sure that Maxwell's equations were valid for uniformly moving charges – that is, constant currents, which produce static magnetic fields. But what about all of the accelerated charges that produce all of the electromagnetic waves, which we had been studying for so long? We were convinced that Maxwell's equations weren't the final word, and that they must be an approximation for something more general. We had some speculative ideas, which we wrote up and submitted as a paper to the American Journal of Physics in 1970. The paper was rejected on the basis, not surprisingly, of being speculative.

I told the story of the paper rejection to a colleague of mine at Oakland University, Bob Edgerton, who had been a faculty member at Dartmouth College. He told me, "I had a colleague at Dartmouth who had a similar experience." The person's name was Miles V. Hayes and he was then an Associate Professor of Engineering at Dartmouth. He had come up with a unified field theory, which was a single equation that he claimed could explain everything. He had submitted three papers on the theory to the Physical Review, all of which were rejected on the grounds that the theory was "speculative." So in 1964, Miles Hayes decided to publish the theory himself as a small book – 70 pages long. He published 400 copies and listed in the back of the book everyone who received one of the 400 copies. He sent 33 copies to the leading physicists of the day, including de Broglie, Heisenberg, and

Dirac. He sent 166 copies to the Chairs of the Departments of Physics of major universities around the world. He sent 89 copies to various libraries around the world, 15 copies to various publishing houses, 47 copies to a variety of other people, and he gave 39 copies to members of the faculty at Dartmouth, including one to Bob Edgerton. Each book was numbered and Bob's was number 309. Miles Hayes died in 1995.

When Bob told me this story he said, "I have the book," and he brought it to me. On the front of the cover jacket, it read, "The universe consists of a complex quaternionic field which is a function of space-time such that its rate of change is proportional to the square of its magnitude." I opened the book and saw the field equation followed by a lot of quaternion algebra. At the time, I didn't know much about quaternions, (many years later, I would teach quaternions in the context of 3-dimensional rotations in computer graphics) so I closed the book, put it on my bookshelf, where it remained for 40 years.

Around 2010, I was going through my books on my bookshelf and pulled out the Miles Hayes book on *A Unified Field Theory*. This time, I read it cover to cover and found it fascinating. I searched on Amazon and found two used copies for sale through obscure small bookshops. My colleague at Oakland University bought one, and my son bought the other. I have never seen a copy available since. There is almost no reference to this book anywhere on the web. It is clear from all of the review articles that have been written about unified field theories, that almost no one knows this book exists. I have never seen it referenced anywhere. And yet, to me, it makes more sense than any of the unified field theories, or theories of everything, that are out there.

To make Hayes' unified field theory more accessible, I will present the entire theory in Chapter 6. The basics of quaternions are given in Appendix A and quaternions are used in Section 2.10 to give a quaternion representation of the Lorentz transformation and also in Section 4.7 to give a quaternion representation of Maxwell's equations. So by the time you get to Chapter 6, quaternions should not be such a mystery.

Returning now to Hamilton, who had just discovered quaternions in 1843 and perhaps, assuming Einstein lived 100 years earlier, may have just derived Maxwell's equations from Coulomb's law and special relativity. Like Carl and I, he may have wondered how Maxwell's equations could be correct for accelerating charges. He would probably think that Maxwell's equations were an approximation to some more general equation. But he would have Maxwell's equations as a single quaternion equation with a complex quaternion containing both the electric and magnetic fields on the left-hand side of the equation, and the charge and current densities on the right-hand side. Perhaps, like Hayes 100 years later, he would be seeking a theory of the universe that was both unified and simple. He would probably agree with Einstein that "there will be no place in the new physics for both fields and matter, that is, particles, because fields will be the only reality." Hamilton might also reason, as did both Einstein and Hayes, that particles should not be defined as the source of the field, but rather very small regions of space in which the field values are very high. In fact, Hayes postulated that particles are standing light waves. But he realized at once that Maxwell's equations were linear and therefore could not account for a stable standing wave of finite size in the absence of material boundaries. For this, and other reasons, Hayes recognized that a single unified field equation must be nonlinear. Perhaps reasoning in a similar way, Hamilton might have come up with the same quaternion field equation as Hayes, in which the right-hand side is a quadratic function of a ψ-field made up of the electric and magnetic fields. As we will see in Chapter 6, if the right-hand side is set to zero, the field equation reduces to Maxwell's equations in free space. If the right-hand side is set to a constant, the field equation reduces to Maxwell's equations with charges and currents. If the right-hand side is set to a linear function of the ψ-field, the field equation reduces to Dirac's equation of quantum mechanics (properly interpreted). This means the equations of classical electromagnetics and quantum mechanics are both approximations of the quadratic field equation of Hayes. This quadratic field equation is also Lorentz invariant – a requirement of an equation which purports to be a theory of everything.

Let's assume that Hamilton did, in fact, come up with the quadratic quaternion field equation 100 years before Hayes did. This equation expands into eight coupled non-linear partial differential equations. We still don't know how to solve these equations analytically (only numerically), but perhaps Hamilton, as the leading mathematician of his day might have come up with solutions. Suppose that he found that one solution is a Lorentz-invariant spectrum of fluctuating electromagnetic radiation (electromagnetic zero-point radiation) in the universe. If so, someone could have derived the blackbody radiation spectrum without any quantum assumptions 100 years before Timothy Boyer did (see Phys. Rev., vol. 182, no. 5, 1969, pp. 1374-1383), and decades before Max Plank derived the blackbody radiation spectrum by introducing the quantum idea. H. E. Puthoff has shown that the ground state of hydrogen can be modeled as a zero-point-fluctuation-determined state without resort to quantum mechanics (see Phys. Rev. D, vol. 35, no. 10, 1987, pp. 3266-3269). Perhaps Hamilton would have found that exact solutions of the nonlinear field equation predict elementary particles as discrete, stable, oscillatory limit cycles in the nonlinear field as suggested by Hayes. Perhaps if these derivations had occurred before Plank, Heisenberg, and Schrödinger, the development of quantum mechanics may have taken a different route.

What else might Hamilton have inferred from the Hayes field equation? Hayes states: "The theory predicts that light waves interact with light waves, or photons with photons. The frequency, amplitude, phase, and other characteristics of light waves are modified by interaction with the other light waves through which they pass in coming from the stars. The greater the distance travelled the greater the modification." Had Hamilton solved this problem and found that the field equation predicts that light from distant stars are red-shifted, then when such observations were actually made, perhaps everyone wouldn't have rushed out and blamed it on the Doppler effect. Perhaps the universe isn't expanding after all. Perhaps the big bang never occurred. The consequences of possibly going down the wrong path for decades and decades boggle the mind.

What about gravity? Hayes goes into some detail about how gravitational fields are ψ-fields, i.e. electromagnetic fields, between neutral particles. If a solution of the field equation is the electromagnetic zero-point radiation described above, then H. E. Puthoff has shown that gravity can be modeled as a zero-point-fluctuation force (see Phys. Rev. A, vol. 39, no. 5, 1989, pp. 2333-3242).

To bring attention to the long-ignored and unknown Hayes book, I have written a novel called *Peggy's Discovery*. In this novel,

Peggy, a high-school senior, with the encouragement of her engineering professor uncle, enters college on a quest to understand the nature of physical reality by challenging conventional wisdom. In the process, she starts a company that revolutionizes higher education, leading her to uncover the secret to a theory of everything. In this novel, Peggy watches her uncle derive Maxwell's equations from Coulomb's law and special relativity just as we do in Chapter 4 of this book. After going through this derivation, Peggy realizes that the exact same derivation can be applied to Newton's law of gravitation and special relativity, meaning that this would lead to equations identical with Maxwell's equations except that the mass density would replace the charge density. She then realizes that there is a second gravitational field, analogous to the magnetic field, which exerts a force only on moving bodies, that this force adds to the normal gravitational force acting on the outer stars of a galaxy and may therefore explain the higher velocities of these stars, meaning there would then be no need to invent dark matter. We discuss Peggy's new theory of gravitation in Chapter 5. As Peggy points out in the novel, this would be the simpler and more obvious way to make Newton's law of gravitation consistent with special relativity, rather than the much more complicated approach that Einstein took in his general theory of relativity, complete with curved four-dimensional space. Perhaps general relativity could have been by-passed altogether as being unnecessary. In fact, Peggy's uncle tells her that Oliver Heaviside, in 1893, twelve years before Einstein published his special theory of relativity, suggested that gravity might be explained with two gravitational fields satisfying equations directly analogous to Maxwell's equations of electromagnetics. Now Peggy has derived these gravitational equations from first principles.

As we will see in this book, Einstein's special theory of relativity is central to understanding electromagnetics; and electromagnetics may be central to a simple unified field theory. Might it be that Einstein's general theory of relativity, with its emphasis on gravity, was a wrong turn down a long path, which has not been successful in leading to a simple unified field theory?

It is too bad Einstein wasn't born 100 years earlier!

Richard E. Haskell

Table of Contents

Chapter 1

Space and Time

1.1 Reference Frames

We are familiar with using coordinate axes as reference frames to describe the motion of a particle. Thus, in Fig. 1.1, we plot the displacement x of the particle as a function of the time t. We say that any point on this graph represents an event. Thus, E_1 is an event which occurs at x_1 at the time t_1 and E_2 is an event which occurs at x_2 at the time t_2. The *world-line* of a particle is the locus of events in the space-time (x-t) graph of Fig. 1.1. The velocity of the particle in Fig. 1.1 is given by $v = \Delta x / \Delta t = \tan \phi$ and is the slope of the world-line.

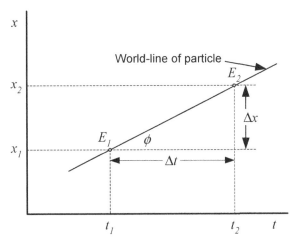

Fig. 1.1 Definition of a world-line

Now we may ask ourselves the following question: Given the event E_1, is the time t_1 found by dropping a perpendicular to the t-axis or is it found by moving parallel to the x-axis? In Fig. 1, it is clear that these two operations are the same so the question may seem unimportant. However, there is no reason, other than convenience, that our coordinate axes x and t should be orthogonal or perpendicular to each other. For example, we could just as well draw them at an oblique angle as shown in Figs. 1.2 and 1.3.

We now see that we have a choice of how to define our components of the events E_1 and E_2. In Fig. 1.2 we move parallel to the coordinate axes while in Fig. 1.3 we drop perpendiculars to the coordinate axes. It is clear that either one of the methods is acceptable. We can therefore pick the one that is most convenient for any particular purpose. In Figs. 1.2 and 1.3 the velocity of the particle is still given by $v - \Delta x / \Delta t$, but note that in neither case is this equal to $\tan \phi$ as it was in Fig. 1. You may wish to find expressions for the velocity of the particle in Figs. 1.2 and 1.3 in terms of ϕ and the angle that the x-axis makes with the t-axis.

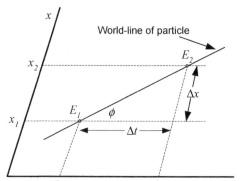

Fig. 1.2 ′Moving parallel to the axes t

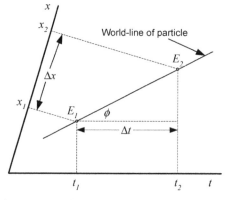

Fig. 1.3 Moving perpendicular to the axes

1.2 The Galilean Transformation

We should like to be able to describe how a given event or a series of events appears to two different observers who move at a constant velocity relative to each other. For example, suppose a bird is flying past a moving train in the direction in which the train is moving. How does the motion of the bird appear to an observer on the train and how does it appear to an observer on the ground?

In order to answer this and similar questions we would like to draw a set of space-time axes (x-t) for each of the observers (the train and the earth) in such a way that they could be used to describe a single event (the bird). Let us see how we might be able to do this.

In Fig. 1.1, the world-line shown is that of a particle moving with a constant velocity relative to an observer at some fixed value of x. To make these ideas concrete let x-t be the space-time coordinate axes for an observer at rest with respect to the earth. The world-line of such an observer would be a horizontal line as shown in Fig. 1.4, since as time goes on he simply remains at the same value of x, namely x_1.

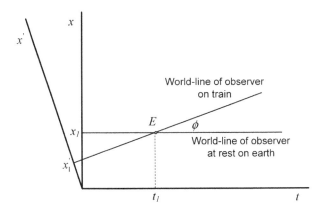

Fig. 1.4 World-lines of two observers

Now we know from Fig. 1.1 that the world-line of a train which moves relative to the earth with a velocity $v = \Delta x/\Delta t = \tan \phi$ will be a straight line inclined at an angle ϕ as shown in Fig. 1.4. We want to draw a set of space-time axes (called x'–t) for an observer at rest on the train. The coordinate axis x' must be such that each point on the world-line of the train has the same value of x'. This will be the case if we draw x' as shown in Fig. 1.4 where we have adopted the convention of Fig. 1.3 and locate the coordinate of an event by dropping perpendiculars to the coordinate axes. Since $x'-t$ are coordinates fixed to the train, as time goes by, an observer on the train remains at the same value of x', namely x_1'. The event E, which occurs when the two observers meet, is represented by the crossing of the two world-lines. Thus, in Fig. 1.4 the observer on the train is just passing by the observer on the ground at time t_1.

It is not yet clear how the scale of distance on the x-axis compares with the scale of distance on the x'-axis. How can we find out? We need some definition of equivalent lengths. Let us denote the system of space coordinates fixed on the earth by S and those fixed on the train by S'. Then let F' be the value of x' at the front of the train and let B' be the value of x' at the back of the train. Then $\Delta x' = F' - B'$ is the length of the train as measured in S' as shown in Fig. 1.5.

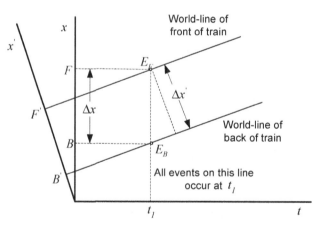

Fig. 1.5 World-lines of the front and back of a train

We know what we mean by the length of the train in S' since the person on the train can leisurely measure the length by stretching a steel tape from the back of the train to the front. But how can the observer on the ground measure the length of the train? One way might be to have two assistants, one stationed at the front of the train and the other stationed at the back of the train. Each assistant has an identical pistol and, at precisely the same time, they each fire a bullet straight down into the ground. The observer on the ground then leisurely walks over and measures the distance between the bullet holes in the ground with a steel tape. The observer on the ground then says that this distance is equal to the length of the train. Where is this length in Fig. 1.5?

The important idea is that the bullets must be fired simultaneously. Suppose they are both fired at time t_1. Then in Fig. 1.5 the event E_B = *firing the bullet at the back of the train* while the event E_F = *firing the bullet at the front of the train*. In the frame of reference fixed to the ground these events are separated by a distance $\Delta x = F - B$. This distance is what the observer on the ground calls the length of the train.

Now if the two observers are to measure the same length for the train (which seems reasonable) we must have $\Delta x = \Delta x'$ in Fig. 1.5. Since these lengths are clearly not equal in Fig. 1.5 we must assign different scales of length to the x and x' axes. This would be cumbersome and we would rather not do it if we can help it. But there is another way out. Since we're finding our coordinates of events by dropping perpendiculars to the coordinate axes, we can make the length of Δx equal to the length of $\Delta x'$ in Fig. 1.5 by tilting the x-axis to the right by the same amount as the x'-axis is already tilted to the left as shown in Fig. 6. It is a simple matter to show that $\Delta x = \Delta x'$ in Fig. 1.6. Thus in Fig. 1.6 the scale lengths

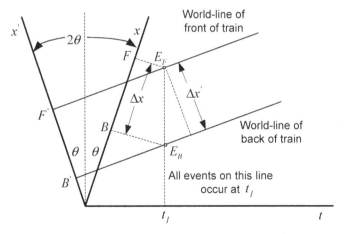

Fig. 1.6 Symmetric tilting of both x and x' axes

along the x' and x axes are the same.

Now since we have tilted the x-axis in Fig. 1.6, it is clear that the velocity of S' (the observer on the train) relative to S (the observer on the ground) will not be $\tan\phi$ as it was in Fig. 1.4. In order to find out what this relative velocity is let's draw the world line of an observer at rest on the train as shown in Fig. 1.7. Note that angle $x_1'E_2x_2$ is equal to 2θ because $x_1'E_2$ is perpendicular to $0x'$ and E_2x_2 is perpendicular to $0x$. Similarly, angle E_2E_1A is equal to θ.

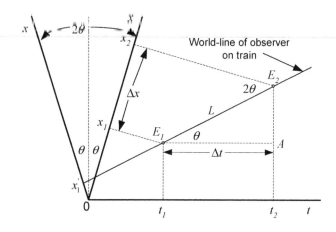

Fig. 1.7 Calculating the relative speed of the train

During the time interval $\Delta t = t_2 - t_1$ the event describing the observer at rest on the train moves from E_1 to E_2. In this time interval, the position of the train as measured by the observer on the ground changes by an amount $\Delta x = x_2 - x_1$. Therefore, the velocity of the train (S') relative to the ground (S) is $u = \Delta x/\Delta t$. From Fig. 1.7

$$\Delta x = L\sin 2\theta$$
$$\Delta t = L\cos\theta$$

and since $\sin 2\theta = 2\sin\theta\cos\theta$ we obtain for the relative velocity u

$$u = \frac{\Delta x}{\Delta t} = \frac{L\sin 2\theta}{L\cos\theta}$$

$$= \frac{2\sin\theta\cos\theta}{\cos\theta}$$

or

$$u = 2\sin\theta \tag{1.1}$$

Figs. 1.6 and 1.7 show how given events appear to two different observers who are moving relative to each other with a constant velocity. The relationship between the coordinates of the two systems shown in these figures is called the *Galilean transformation*. As an example of this transformation we will now look at how velocities transform from one system to the other.

1.3 Velocity Transformation

Let us now consider the problem of a bird flying past the moving train. Since the bird is flying faster than the train its world line must have a positive slope relative to the world line of the moving train. The situation is shown in Fig. 1.8.

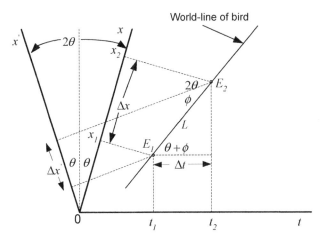

Fig. 1.8 Calculating the velocity of the bird

From Eq. (1.1) we know that u = velocity of S' (train) relative to S (ground)

$$u = 2\sin\theta$$

Also let

$$v = \text{velocity of bird relative to } S = \Delta x/\Delta t$$

$$v' = \text{velocity of bird relative to } S' = \Delta x'/\Delta t$$

From Fig. 1.8

$$\Delta x = L \sin (2\theta + \phi)$$
$$\Delta x' = L \sin \phi$$
$$\Delta t = L \cos (\theta + \phi)$$

We can therefore write

$$\Delta x = L \sin (2\theta + \phi)$$
$$= L (\sin 2\theta \cos \phi + \cos 2\theta \sin \phi)$$
$$= L \left[2 \sin \theta \cos \theta \cos \phi + (1 - 2 \sin^2 \theta) \sin \phi \right]$$
$$= L \sin \phi + 2 \sin \theta L (\cos \theta \cos \phi - \sin \theta \sin \phi)$$
$$= L \sin \phi + 2 \sin \theta L \cos (\theta + \phi)$$

or

$$\Delta x = \Delta x' + u \Delta t \tag{1.2}$$

which is the equation for the Galilean transformation. It follows that

$$\frac{\Delta x}{\Delta t} = \frac{\Delta x'}{\Delta t} + u$$

or

$$v = v' + u \tag{1.3}$$

This is, of course, just what you would expect. The velocity of the bird relative to the ground is the velocity of the bird relative to the train plus the velocity of the train relative to the ground. Why all the fuss? You say you could have figured that out in your head much easier. And you're right. If this were the way nature works, it certainly would not be worth all the effort to make all these graphs. But this is not the way nature works! We've made a mistake somewhere in our thinking. The Galilean transformation does not predict everything that we observe experimentally. What then is the proper picture of space and time? We will see in the next chapter that the graphs we have constructed for the Galilean transformation must be modified. However, our understanding of the use of these graphs will help us considerably in understanding this new theory of relativity. We will find that it is time that is the culprit and that the proper picture of space and time is one in which the asymmetry of Figs. 1.6-1.8 disappears and space and time are put on an equal basis.

Chapter 2

Relativistic Kinematics

2.1 The Principle of Relativity

Newton's laws of motion are stated as holding only with respect to an inertial frame of reference. Such a frame of reference is defined by Newton's first law. That is, an inertial frame of reference is one in which the law of inertia holds. It follows that there are an infinite number of systems of reference (inertial systems) moving uniformly and rectilinearly with respect to each other in which the law of motion ($F = ma$) is of the identical form.

The preceding is a statement of the *principle of relativity* for classical mechanics. It asserts that absolute uniform motion cannot be detected by any experiment of classical mechanics. For example, if you were riding along on a train moving with uniform velocity, there is no mechanical experiment that you can do completely within the train that could tell you whether or not the train was moving. As far as you could tell the train might very well be at rest. This is certainly consistent with our everyday experience. Of course if the train accelerates or decelerates we can feel or detect this motion but as long as it is moving uniformly we have no way of detecting the motion.

Now this is a very satisfying principle of physics. It says that it is unnecessary to assume a special or specific frame of reference with respect to which the laws of mechanics are to hold. It says that any of an infinite number of (inertial) frames of reference is equally acceptable and that no one frame of reference is given a special position of importance above any other frame. If this weren't so, then we would have to write a different equation of motion for every different frame of reference. It is clearly desirable to have the laws of physics be independent of the particular frame of reference from which the experiment is observed.

We see that for classical mechanics things seem to be in good shape with respect to the principle of relativity. What about the rest of physics? Toward the end of the last century it appeared as though the laws of electromagnetism violated the principle of relativity and as a result physicists thought that it would be possible to detect absolute uniform motion by certain experiments involving the propagation of light, which is a form of electromagnetic energy. Many such experiments were carried out, the most famous of which is the Michelson-Morley experiment, which was designed to detect the uniform motion of the earth relative to a hypothetical ether, assumed to be at rest throughout absolute space. However, this experiment was unable to detect the absolute uniform motion of the earth.

There were many other examples of electromagnetic phenomena which were inconsistent with the concept of absolute rest. As a result, in 1905 Albert Einstein took as his first postulate that the principle of relativity that we stated above as holding for the laws of mechanics also holds for the laws of electrodynamics and therefore optics. We can therefore state Einstein's first postulate which is called the *Principle of Relativity* as follows: *Absolute uniform motion cannot be detected by any means.* This is to say that the concept of absolute rest and the ether have no meaning.

2.2 The Nature of Light

The Principle of Relativity described in the previous section does not seem to be a particularly disturbing postulate. As a matter of fact it seems quite reasonable. However, when Einstein stated this principle as his first postulate it did not seem so reasonable. This was because Maxwell's equations of electromagnetism predicted that light would travel with a constant velocity c. The question is – a velocity c with respect to what? It was thus supposed that it must be with respect to an ether, which was at absolute rest in the universe. It then followed from the Galilean transformation that absolute uniform motion with respect to the ether could be detected. As pointed out above, all attempts to detect such motion have failed.

In addition to his first postulate of the Principle of Relativity, Einstein stated a second postulate concerning the nature of light. It was that *light is propagated in empty space with a velocity c, which is independent of the motion of the source.*

We know that the velocity of some things do depend on the velocity of the source. For example, the velocity of a bullet will appear to travel faster to an observer on the ground if the gun is moving in the direction in which it is fired. On the other hand, the velocity of sound does not depend on the velocity of the source but always has the same velocity with respect to the air. Physicists at the end of the last century thought light must act the same way. They believed that the velocity of light was independent of the velocity of the source. After all, the velocity of light must have the same value with respect to the ether. Thus Einstein's second postulate would seem quite natural.

However, we have seen that Einstein's first postulate implies that there is no ether. Thus, at first sight, there seems to be no way that both postulates can be true. Einstein showed that in order for both postulates to be true we must modify our ideas about the nature of time. Let us remind ourselves that the reason we accept the two postulates by themselves is that they agree with our experience. The combination of the two postulates leads to predictions which, at first sight, seem quite unlikely. However, many experiments have subsequently shown that these unlikely events do, in fact, occur.

2.3 The Nature of Time

In order to understand the dilemma of Einstein's two postulates consider the Galilean transformation represented by Fig. 1.8. The velocity of the bird is different when viewed from two different inertial frames (the ground and the train). However, Einstein's postulates state that if we send a light signal from the back of the train to the front of the train then an observer on the train and an observer on the ground must both measure the velocity of the light beam to be c. How can we draw a world-line for the light beam in Fig. 1.8 so this will be so? Only a vertical world-line would have the same velocity in both reference frames but this velocity would be infinite and not c. What else can we do?

The only way that Fig. 1.8 can be modified so the velocity of light will have the same value c in both reference frames is to split the t-axis as shown in Fig. 2.1. If we label the time axes in units of ct and ct', then the world-line of a light signal must have a slope of

+1 in both frames of reference. We notice that by splitting the time axis, this is accomplished, and in each reference frame the velocity of light is given by $\Delta x/\Delta t = \Delta x'/\Delta t' = c$.

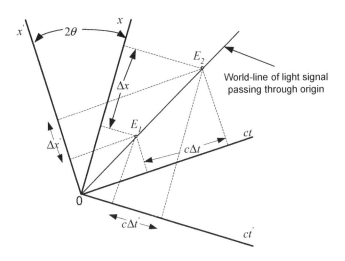

Fig. 2.1 Splitting the t-axis

   This splitting of the time axis is the central result of relativity theory. We must try to understand what it means. Let us first determine what the relative velocity between the reference frames S and S' is in terms of our new picture of space-time given by Fig. 2.1. A particle at rest in S' will have a zero velocity relative to S'. The world-line of such a particle is shown in Fig. 2.2. The velocity of S' relative to S which we denote by u is then the velocity of the particle at rest in S' relative to S. From Fig. 2.2 this velocity will be $u = \Delta x/\Delta t$. Note that we have let $2\theta = \alpha$ and also that the world-line of the particle at rest in S' is parallel to the ct-axis. This is because the axes x' and ct are perpendicular as are the axes x and ct'. From Fig. 2.2, we therefore can write

$$\Delta x = L\sin\alpha$$
$$c\Delta t = L$$

so that

$$u = \frac{\Delta x}{\Delta t} = c\sin\alpha$$

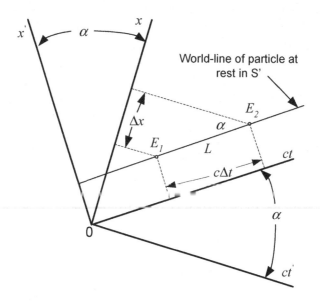

Fig. 2.2 Finding the relative velocity between S and S'

or

$$\sin \alpha = u/c \tag{2.1}$$

Since $\sin^2 \alpha + \cos^2 \alpha = 1$ it follows that

$$\cos \alpha = \sqrt{1 - \sin^2 \alpha}$$
$$\cos \alpha = \sqrt{1 - \frac{u^2}{c^2}} \tag{2.2}$$

We immediately see from Eq. (2.1) that u must be an appreciable fraction of the speed of light c in order for α to have a significantly large value. For example, if $u = 0.1c$, then $\alpha = 5.7°$. The angle α increases as u increases and approaches $90°$ as u approaches c.

To understand the nature of time as depicted by Figure 2.2, let us consider how two observers on a train, one at the front and one at the back, might synchronize their clocks so they will know they read the same time. One way might be for the two persons to meet somewhere on the train and set their clocks to the same time. Then one person moves to the back of the train and the other to the front of the train. Can you think of anything wrong with this method? How do you know that the beating of the clocks remained the same when they were in relative motion? The only direct comparisons of clocks that we can make are when they are at the same place at the same time. Probably the worst thing about the above method of synchronizing clocks from the physicists point of view is that there is no way of testing whether the clocks remained synchronized except by sending time signals once they are at the front and back of the train. This suggests that a better method of synchronizing clocks would be to use time signals to begin with. This could be done in the following way.

We measure the length of the train with a steel tape and then place a third person at exactly the center of the train. At a certain time that person explodes a flashbulb which sends a light signal in both directions at the constant velocity c. Each person at the front and back of the train has a clock which automatically starts when the light signal arrives. Now since they know (by Einstein's postulates) that the light will take the same time to travel to the front of the train as it does to travel to the back, they are justified in saying that the signals will arrive at the front and back of the train simultaneously and that their clocks will be synchronized. Let us see what this looks like in our figures. Fig. 2.3 shows the world-lines of the front (F'), back (B'), and center (C') of the train. The event E_1 is the exploding of the flashbulb at the center of the train at time t_1'. Event E_2 is the arrival of the light signal at the back of the train and event E_3 is the arrival of the light signal at the front of the train. As advertised these events occur simultaneously to the observer on the train at the time t_2'.

Now how does an observer on the ground in reference frame S describe what is going on? We see immediately from Fig. 2.3 that event E_1 occurs at time t_1, event E_2 occurs at time t_2, and event E_3 occurs at time t_3, all of which are different. We therefore see that events E_2 and E_3, which were simultaneous on the train, are not simultaneous when viewed from the ground. Why is this so? Remember that Einstein's postulates say that both observers must measure c for the velocity of light and that it is independent of the velocity of the source. Thus, the observer on the ground must

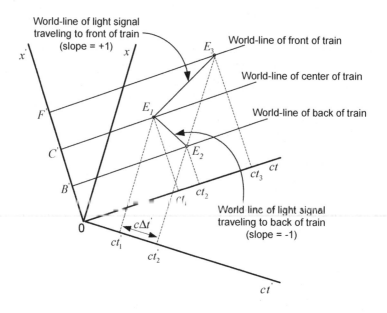

Fig. 2.3 Showing that simultaneity is relative

measure the light signal to travel at the velocity c in both directions. But since the train is moving with a velocity u with respect to the ground the light signal will clearly arrive at the back of the train, which is moving into the light signal, before it arrives at the front of the train, which is moving away from the light signal. This is exactly what an observer on the ground sees as is shown in Fig. 2.3.

Both observers are equally correct in describing the events in Fig. 2.3. All that it means is that the concept of simultaneity is a relative concept. It depends on your frame of reference. *There is no such thing as absolute simultaneity.* If this seems strange to you, note that there is now a certain symmetry of space and time as shown in Fig. 2.3. It is not strange to you that two events which occur at the same place to one observer don't occur at the same place to another observer. All the way back in Fig. 1.7 the events E_1 and E_2 occur at the same place on the train but are separated by the distance Δx on the ground. The important thing is that they occur at different times. In a similar way, the events E_2 and E_3, which occur at the same time on the train in Fig. 2.3, are separated in time on the ground by the time interval $c(t_3 - t_2)$. Again the important thing is that these events now occur at different locations.

2.4 Time Dilation

Suppose Carl is in the caboose at the back of the train and has a clock which he uses to keep time. As he passes Alice (A), who is on the ground and has her own clock, a photograph is taken which shows both clocks. They happen to read the same value. Now Betsy (B) is a certain distance down the track, and has a clock that has previously been synchronized with Alice's clock by using time signals as explained in the previous section. As the caboose passes Betsy, another photo is taken which shows both Betsy's clock and Carl's clock in the caboose. Carl's clock reads less than Betsy's clock. What has happened?

Let us follow this sequence of events on our diagrams. The world-line of Carl in the caboose is shown in Fig. 2.4. Event E_1 is the caboose passing Alice (A) and event E_2 is the caboose passing Betsy (B). The time between these two events is measured to be $\Delta t'$ to Carl on the train, while it is measured to be Δt to Alice and Betsy on the ground. From Fig. 2.4 we see that

$$\Delta t' = \Delta t \cos \alpha$$

or, from Eq. (2.2)

$$\Delta t = \frac{\Delta t'}{\cos \alpha} = \frac{\Delta t'}{\sqrt{1 - \frac{u^2}{c^2}}} \tag{2.3}$$

Therefore Δt is greater than $\Delta t'$. How can we understand this time dilation of the time interval between events E_1 and E_2 as measured by observers on the ground?

First let us review exactly what has happened. The clock in the primed system (i.e., the clock in the caboose) has moved from A to B and has recorded the time interval $\Delta t'$. This time interval is measured by one and the same clock. Such time, which is measured by a single clock, is called *proper time*. On the other hand, it required two different clocks which were separated in space to

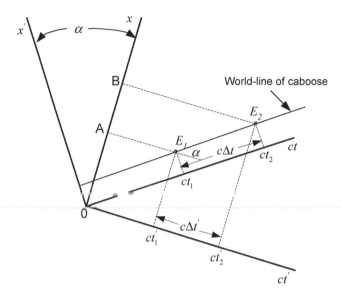

Fig. 2.4 World-line of caboose

measure the time interval Δt.   This kind of time is called non-proper, or *coordinate time*.   It is always the case that the shortest time interval is shown by the clock which measures proper time. Let us try to uncover the source of this dilation.

In order to measure the time interval Δt on the ground we had to synchronize the clocks at A and B. We found previously that we could do this by exploding a flashbulb at a point C exactly half-way between A and B. If the clocks at A and B both start automatically just as the light signal reaches each one, then we say that the clocks at A and B are synchronized. Let's suppose that the light signal reaches A (and therefore B) just as the caboose is passing A. This situation is shown in Fig. 2.5. Again, the event E_1 is the caboose passing A while the event E_2 is the caboose passing B as in Fig. 2.4. The event E_0 is the exploding of the flashbulb at C. The arrival of this signal at B is event E_3 while the arrival of this signal at A is E_1. Events E_1 and E_3 occur simultaneously at time t_1 in the reference frame of the earth.

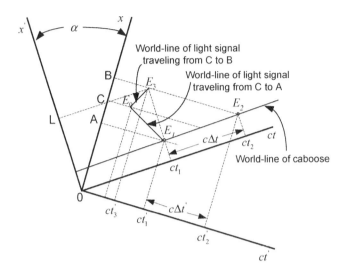

Fig. 2.5 Understanding time dilation

However, as we already know, these events are not simultaneous in the reference frame of the moving caboose. In this frame of reference, the light signal arrives at B (at time t_3') before it arrives at A (at time t_1'). Therefore, from the point of view of Carl in the caboose, Betsy (B) starts her clock before she "ought to", i.e., before the caboose reaches Alice (A) at time t_1'. Therefore, it is not surprising that when Alice and Betsy compare their clocks, they get a longer time interval between events E_1 and E_2 than does Carl in the caboose.

However, it's not quite that simple. For if Carl in the caboose tried to correct this "error" by starting his clock at t_3' instead of t_1' he would find that $t_2'-t_3'$ is *greater* than $\Delta t = t_2 - t_1$ by the same amount that Δt was previously greater than $\Delta t' = t_2'-t_1'$. Why? Because Δt is now the proper time since it measures the time between events E_3 and E_2 with the single clock at B. However, the time t_3' could only be determined by someone on the train who happened to be passing B at the instant it was receiving the light signal from C.

Perhaps Larry (L) in the locomotive at the front of the train just passes Betsy (B) at E_3 as shown in Fig. 2.5. Larry can note the time t_3' on his clock and later compare it with Carl in the caboose who measures t_2' when passing Betsy (B). But remember, these two clocks must have previously been synchronized, but of course they will not appear synchronized to observers on the ground. Thus, the time interval $t_2' - t_3'$ is an improper time interval since it requires two different clocks (one in the locomotive and one in the caboose) to measure the time interval. It is, therefore, longer than the corresponding proper time interval, Δt.

We therefore see that time dilation is a direct consequence of the fact that simultaneity is only a relative concept. This, in turn, is a direct consequence of the fact that the velocity of light measures the same in all inertial frames. As a result, it is impossible to synchronize clocks which are in relative motion.

2.5 Length Contraction

Closely associated with the idea of time dilation is the phenomenon of length contraction. Suppose the people on the train measure the length of the train with a steel tape and find it to be $\Delta x' = F' - B'$ as shown in Fig. 2.6. How can observers on the ground determine the length of the train? You will recall in Section 1.2 that this length was measured by firing bullets simultaneously from the front and back of the train into the ground and then measuring the distance between the bullet holes. But that was before we knew that simultaneity is only relative. Events which appear to be simultaneous on the train will not be simultaneous to observers on the ground. What we generally mean when we talk about the length of the train as measured by observers on the ground is the distance between two observers on the ground, one who is beside the front of the train, and the second who is simultaneous beside the back of the train. This is, of course, simultaneous in the reference frame of the ground. Therefore, we cannot use the method of firing bullets simultaneously from the train. What can we do?

We can have Betsy on the ground who notes the time on her clock t_1 as the back of the train passes. We label this event E_1 and designate the location of Betsy by B. Now where is the front of the

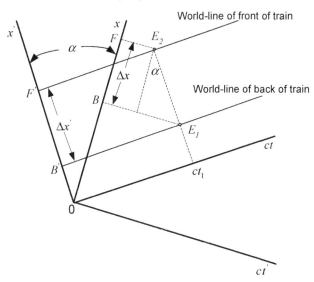

Fig. 2.6 Understanding length contraction

train at this same time t_1? We don't know ahead of time, so we must station observers all along the track, each being a known distance from Betsy (B) and having a clock, previously synchronized with Betsy's clock. Then each one records the time at which the front of the train passes by. They later get together and compare notes. One of these many observers will have recorded the same time t_1 that Betsy recorded when the back of the train passed by B.

This observer will be Frank, who was at the location F, and we define the length of the train as measured on the ground to be $\Delta x = F - B$ as shown in Fig. 2.6. From this figure we immediately see that

$$\Delta x = \Delta x' \cos \alpha$$

which, using Eq. (2.2), can be written as

$$\Delta x = \Delta x' \sqrt{1 - \frac{u^2}{c^2}} \tag{2.4}$$

Thus, the length of the train is shorter when measured by observers on the ground than it is when measured in the train. This result is clearly another consequence of relative simultaneity and is closely related to time dilation.

Now Δx and $\Delta x'$ are lengths measured in the direction of the relative motion u. What about lengths such as Δy and $\Delta y'$ measured perpendicular to the relative motion. It should be apparent that these lengths will be the same. To see this, consider Fig. 2.7 in which a light signal is sent out from point C just as C' (the center of the front of the train) passes C. It line AB is perpendicular to the direction of motion it is clear that A and B will receive the light signal simultaneously in the earth's reference frame. But it is also clear that A' and B' on the train will also receive the light signal simultaneously. Thus, if A' and B' fire bullets into the ground simultaneously (in their frame), this will also be simultaneous in the earth's frame, so that if A and B go over and measure the distance between the holes in the ground, they will measure the same width w of the train as A' and B' measure on the train. Thus, if the relative motion u is in the x-direction, then $\Delta y' = \Delta y$ and $\Delta z' = \Delta z$.

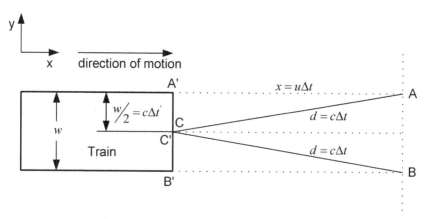

Fig. 2.7 Lengths perpendicular to the direction
of motion remain unchanged

Of course, time intervals are still measured differently in the two reference frames. Thus, although A and B receive the light signals simultaneously and so do A' and B', the two sets of observers differ on how long they say it takes the light to reach them from the source since both must measure the velocity of light to be c. From Fig. 2.7 we see that to observers on the ground

$$\frac{w}{2} = \sqrt{d^2 - x^2} = \Delta t \sqrt{c^2 - u^2} = c\Delta t \sqrt{1 - \frac{u^2}{c^2}}$$

while on the train $\dfrac{w}{2} = c\Delta t'$. Therefore $\Delta t' = \Delta t \sqrt{1 - \dfrac{u^2}{c^2}}$ just as was found in Section 2.4.

2.6 The Lorentz Transformation

In Sections 1.2 and 1.3 we obtained diagrams which represented the so-called Galilean transformation. This transformation is shown in Fig. 1.8 for the case of a bird flying past a moving train and is given by Eq. (1.3). However, we have now seen that Fig. 1.8 is really not an accurate picture of space and time but must be replaced by Fig. 2.8.

From this figure we can write

$$\Delta x = L \sin(\phi + \alpha) \tag{2.5}$$

$$\Delta x' = L \sin \phi \tag{2.6}$$

$$c\Delta t = L \cos \phi \tag{2.7}$$

$$c\Delta t' = L \cos(\phi + \alpha) \tag{2.8}$$

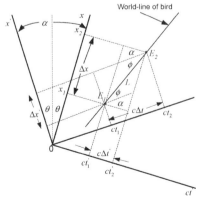

Fig. 2.8 Deriving the Lorentz transformation

From Eqs. (2.1) and (2.2) recall that

$$\sin \alpha = \frac{u}{c} \tag{2.9}$$

and let

$$\gamma = \frac{1}{\cos \alpha} = \frac{1}{\sqrt{1 - \dfrac{u^2}{c^2}}} \tag{2.10}$$

Then, from Eqs. (2.5), (2.6), (2.7) and (2.10), it follows that

$$\begin{aligned}
\Delta x &= L\left(\sin \phi \cos \alpha + \cos \phi \sin \alpha\right) \\
&= \Delta x' \cos \alpha + c\Delta t \sin \alpha \\
&= \frac{\Delta x'}{\gamma} + u\Delta t
\end{aligned}$$

from which

$$\Delta x' = \gamma\left(\Delta x - u\Delta t\right) \tag{2.11}$$

Also, from Eqs. (2.8), (2.7), (2.6), (2.10) and (2.9), it follows that

$$\begin{aligned}
c\Delta t' &= L\left(\cos \phi \cos \alpha - \sin \phi \sin \alpha\right) \\
&= c\Delta t \cos \alpha - \Delta x' \sin \alpha \\
&= \frac{c\Delta t}{\gamma} - \frac{u}{c}\Delta x'
\end{aligned}$$

from which

$$\Delta t = \gamma\left(\Delta t' + \frac{u}{c^2}\Delta x'\right) \tag{2.12}$$

Since there is complete symmetry between the observation in the two frames of reference except that the sign of the relative velocity u changes, we can obtain Δx in terms of $\Delta x'$ and $\Delta t'$ by simply interchanging the primed and unprimed quantities in Eq. (2.11) and changing the sign of u. Thus

$$\Delta x = \gamma\left(\Delta x' + u\Delta t'\right) \tag{2.13}$$

This equation can be verified by substituting Eq. (2.12) in Eq. (2.11) and solving for Δx.

In a similar manner from Eq. (2.12) we can immediately write

$$\Delta t' = \gamma\left(\Delta t - \frac{u}{c^2}\Delta x\right) \tag{2.14}$$

which can also be verified by substituting Eq. (2.11) in (2.12) and solving for $\Delta t'$. Eqs. (2.11) - (2.14) are called the *Lorentz transformation* which can be summarized as follows:

$$\Delta x = \gamma\left(\Delta x' + u\Delta t'\right)$$

$$\Delta t = \gamma\left(\Delta t' + \frac{u}{c^2}\Delta x'\right)$$

$$\Delta x' = \gamma\left(\Delta x - u\Delta t\right) \tag{2.15}$$

$$\Delta t' = \gamma\left(\Delta t - \frac{u}{c^2}\Delta x\right)$$

2.7 Relavistic Velocity Transformation

In Section 1.3 we found that the Galilean transformation gave rise to a velocity transformation of the form (see Eq. (1.3))

$$v = v' + u \tag{2.16}$$

But we see from Eqs. (2.15) that the Lorentz transformation will produce a different velocity transformation law. In particular

$$\frac{\Delta x}{\Delta t} = \frac{\gamma\left(\Delta x' + u\Delta t'\right)}{\gamma\left(\Delta t' + \frac{u}{c^2}\Delta x'\right)}$$

If we divide numerator and denominator by $\Delta t'$, we obtain

$$\frac{\Delta x}{\Delta t} = \frac{\left(\Delta x'/\Delta t' + u\right)}{\left(1 + \dfrac{u\Delta x'}{c^2 \Delta t'}\right)}$$

or

$$v = \frac{v' + u}{1 + \dfrac{uv'}{c^2}} \qquad\qquad (2.17)$$

Similarly, from Eq. (2.15)

$$\frac{\Delta x'}{\Delta t'} = \frac{\gamma\left(\Delta x - u\Delta t\right)}{\gamma\left(\Delta t - \dfrac{u}{c^2}\Delta x\right)} = \frac{\Delta x/\Delta t - u}{1 - \dfrac{u\Delta x}{c^2 \Delta t}}$$

or

$$v' = \frac{v - u}{1 - \dfrac{uv}{c^2}} \qquad\qquad (2.18)$$

Eqs. (2.17) and (2.18) are the velocity transformation laws for the Lorentz transformation. The first thing to notice is that they reduce to the Galilean transformation law, Eq. (2.16), for low relative velocities, $u << c$.

Let's suppose the train has a speed relative to the earth of $u = 0.6c$, a very high speed train! Suppose also that a bird flies past the train with a velocity relative to the train of $v' = 0.8c$, a very fast bird. How fast is the bird flying relative to the ground? Our old Galilean transformation, Eq. (2.16), would tell us that $v = 0.8c + 0.6c = 1.4c$, faster than the speed of light. However, the Lorentz transformation predicts, by Eq. (2.17), that

$$v = \frac{0.8c + 0.6c}{1 + \dfrac{(0.6c)(0.8c)}{c^2}} = \frac{1.4c}{1.48} = 0.946c$$

which is still less than the speed of light.

Let the train be moving with a velocity $u = kc$ ($k < 1$) relative to the earth. A light signal which travels in the train frame of reference with a velocity $v' = c$ will, by Eq. (2.17), travel in the earth frame of reference with a velocity

$$v = \frac{c + kc}{1 + \frac{kc^2}{c^2}} = \frac{1+k}{1+k} c = c$$

as it should by Einstein's postulates.

Now the preceding examples show that Eqs. (2.17) and (2.18) predict that if an object is traveling less than the speed of light in one reference frame it will travel less than the speed of light in all other frames which move relative to the first with velocities less than c. But is it possible for objects to travel faster than the speed of light in the first place? Let us consider what this would have to mean by referring to Fig. 2.9.

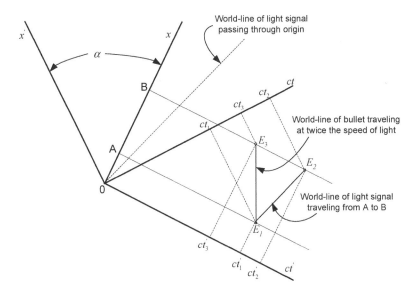

Fig. 2.9 A bullet traveling at twice the speed of light

Suppose in the earth's reference frame a flashbulb is exploded at point A at time t_1. This is event E_1. Event E_2 is the arrival of the light signal at point B at time t_2. An observer on a fast moving train would say that event E_1 occurred at time t_1' and event E_2

occurred at time $t_2{}'$. Now suppose that at the instant the flashbulb exploded, a gun was fired which was able to fire a bullet at twice the speed of light in the earth's frame of reference. It would, therefore, arrive at point B at the time t_3. This is event E_3 in Fig. 2.9. But to the observer on the moving train this event occurs at time $t_3{}'$ which is less than $t_1{}'$. That is, to the observer on the train the bullet arrives at B before it was fired at A!

We might be able to accept this if there were no way for observers in the primed system to prevent the shooting. But if they can have bullets that travel faster than c in their frame, they can prevent the shooting. To see this, consider Fig. 2.10.

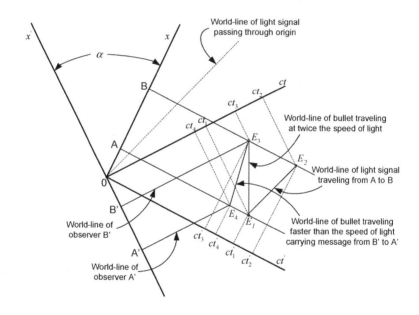

Fig. 2.10
Illustrating logical contradiction when speed of light is exceeded

At time t_1, A fires a bullet (event E_1) at twice the speed of light and kills B at time t_3 (event E_3). Observer B' who is just passing by at time $t_3{}'$ observes the killing and scoops up the dead body of B. Observer B' immediately sends a message on a bullet traveling faster than c to A', telling of the shooting. This message arrives at A' at time $t_4{}'$ (event E_4). However, at this time, A' is just passing

A and, since the time in the reference frame of A is t_4, which is less than t_1, then A hasn't yet fired the shot, so A' disarms A. But B' has B's dead body! To avoid this kind of serious contradiction, we must conclude that the bullets, or anything else, cannot travel faster than the speed of light in any reference frame.

2.8 Vector Representation of the Lorentz Transformation

The Lorentz transformation was found in Section 2.6 to be given by Eqs. (2.15) when the relative velocity u was in the x direction. For this case, we also found that $\Delta y' = \Delta y$ and $\Delta z' = \Delta z$. If the origins of the two coordinate systems coincide at $t = t' = 0$ then we can write the Lorentz transformation as

$$
\begin{aligned}
x' &= \gamma(x - ut) \\
y' &= y \\
z' &= z \\
t' &= \gamma\left(t - \frac{ux}{c^2}\right)
\end{aligned}
\tag{2.19}
$$

$$
\begin{aligned}
x &= \gamma(x' + ut') \\
t &= \gamma\left(t' + \frac{ux'}{c^2}\right)
\end{aligned}
\tag{2.20}
$$

where

$$
\gamma = \frac{1}{\sqrt{1 - \dfrac{u^2}{c^2}}}
$$

Let us consider the general case in which the relative motion can be in an arbitrary direction as shown in Fig. 2.11. The speed of S' relative to S is u and α_i is a unit vector in the direction of motion

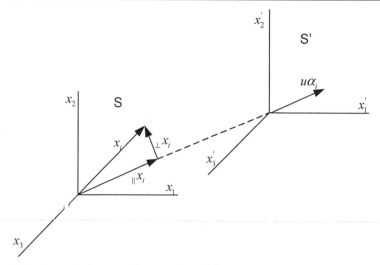

Fig. 2.11 Reference frames for arbitrary direction of motion

$(\alpha_i \alpha_i = 1)$.[1]

Let

$$\beta_i = \alpha_i \frac{u}{c}$$

then

$$\beta^2 = \beta_i \beta_i = \frac{u^2}{c^2}$$

and

$$\gamma = \frac{1}{\sqrt{1 - \beta^2}}$$

If x_i = the displacement vector in S

x_i' = the displacement vector in S'

$_{\parallel}x_i$ = the component of x_i parallel to the direction of motion

$_{\perp}x_i$ = the component of x_i perpendicular to the direction of motion

---

[1] α_i (i =1,3) are the three components of the vector $\boldsymbol{\alpha}$. Repeated indices are summed from 1 to 3; i.e., $\alpha_i \alpha_i = \alpha_1^2 + \alpha_2^2 + \alpha_3^2 = 1$

then

$$_{\parallel}x_i = x_j\alpha_j\alpha_i \tag{2.21}$$

$$_{\perp}x_i = x_i - _{\parallel}x_i \tag{2.22}$$

and similarly for $_{\parallel}x_i{}'$ and $_{\perp}x_i{}'$.

The Lorentz transformation given by Eq. (2.19) can then be written as

$$
\begin{aligned}
_{\parallel}x_i{}' &= \gamma\left(_{\parallel}x_i - u\alpha_i t\right) \\
_{\perp}x_i{}' &= _{\perp}x_i \\
t' &= \gamma\left(t - \frac{u}{c^2}\alpha_j x_j\right)
\end{aligned}
\tag{2.23}
$$

We can therefore write

$$
\begin{aligned}
x_i{}' &= _{\parallel}x_i{}' + _{\perp}x_i{}' \\
&= \gamma\left(_{\parallel}x_i - u\alpha_i t\right) + _{\perp}x_i \\
&= \gamma_{\parallel}x_i - \gamma u\alpha_i t + x_i - _{\parallel}x_i \\
&= (\gamma-1)x_j\alpha_j\alpha_i - \gamma u\alpha_i t + x_i \\
&= \left[\delta_{ij} + (\gamma-1)\alpha_i\alpha_j\right]x_j - \gamma u\alpha_i t
\end{aligned}
$$

In the last step δ_{ij} is the Kronecker delta that is equal to 1 when $i = j$ and is equal to 0 when $i \neq j$; therefore, $\delta_{ij}x_j = \delta_{i1}x_1 + \delta_{i2}x_2 + \delta_{i3}x_3 = x_i$. Thus, the Lorentz transformation can be written in the vector form

$$
\begin{aligned}
x_i{}' &= \left[\delta_{ij} + (\gamma-1)\alpha_i\alpha_j\right]x_j - \gamma u\alpha_i t \\
x_i &= \left[\delta_{ij} + (\gamma-1)\alpha_i\alpha_j\right]x_j{}' + \gamma u\alpha_i t' \\
t' &= \gamma\left(t - \frac{u}{c^2}\alpha_j x_j\right) \\
t &= \gamma\left(t' + \frac{u}{c^2}\alpha_j x_j{}'\right)
\end{aligned}
\tag{2.24}
$$

We can find the general velocity transformation by noting that

$$dx_i' = \left[\delta_{ij} + (\gamma - 1)\alpha_i\alpha_j \right] dx_j - \gamma u \alpha_i dt$$

$$dt' = \gamma \left(dt - \frac{u}{c^2} \alpha_j dx_j \right)$$

Therefore,

$$v_i' = \frac{dx_i'}{dt'} = \frac{\left[\delta_{ij} + (\gamma - 1)\alpha_i\alpha_j \right] v_j - \gamma u \alpha_i}{\gamma \left(1 - \frac{u}{c^2} \alpha_j v_j \right)}$$ (2.25)

and

$$v_i = \frac{dx_i}{dt} = \frac{\left[\delta_{ij} + (\gamma - 1)\alpha_i\alpha_j \right] v_j' + \gamma u \alpha_i}{\gamma \left(1 + \frac{u}{c^2} \alpha_j v_j' \right)}$$ (2.26)

Note that if u is in the x_1 direction; i.e., $\alpha_i = (1,0,0)$ then Eq. (2.25) reduces to

$$v_1' = \frac{\left[1 + (\gamma - 1) \right] v_1 - \gamma u}{\gamma \left(1 - \frac{uv_1}{c^2} \right)} = \frac{v_1 - u}{1 - \frac{uv_1}{c^2}}$$ (2.27)

which agrees with Eq. (2.18). In addition

$$v_2' = \frac{v_2}{\gamma \left(1 - \frac{uv_1}{c^2} \right)} = \frac{v_2 \sqrt{1 - u^2/c^2}}{1 - \frac{uv_1}{c^2}}$$ (2.28)

2.9 4-Vector Representation of the Lorentz Transformation

If a flashbulb goes off at the origin of a fixed reference frame S, a spherical light wave front will expand at the speed of light c. Thus, at time t, this wave front will have moved a distance ct and will be described by the equation of a sphere

$$x^2 + y^2 + z^2 = c^2 t^2 \qquad (2.29)$$

By the fundamental postulate of special relativity, a similar spherical wave front will expand at the same speed of light c in any inertial frame S' moving at a uniform velocity relative to the fixed reference frame S. In the reference frame S', its wave front will also be described by the equation of a sphere

$$x'^2 + y'^2 + z'^2 = c^2 t'^2 \qquad (2.30)$$

Thus we see that $x^2 + y^2 + z^2 - c^2 t^2$ is an invariant under a Lorentz transformation. That is,

$$x^2 + y^2 + z^2 - c^2 t^2 = x'^2 + y'^2 + z'^2 - c^2 t'^2 \qquad (2.31)$$

We can verify this for the case of S' moving in the x-direction with a speed u. Then, Eq. (2.31) reduces to

$$x^2 - ct^2 = x'^2 - ct'^2 \qquad (2.32)$$

From Eq. (2.19), the Lorentz transformation for this case is

$$x' = \gamma(x - ut)$$
$$t' = \gamma\left(t - \frac{ux}{c^2}\right) \qquad (2.33)$$

from which we can write

$$x'^2 - c^2 t'^2 = \gamma^2 (x - ut)^2 - c^2 \gamma^2 \left(t - \frac{ux}{c^2} \right)^2$$

$$= \gamma^2 x^2 - \gamma^2 2xut + \gamma^2 u^2 t^2 - c^2 \gamma^2 t^2 + \gamma^2 2uxt - \gamma^2 \frac{u^2 x^2}{c^2}$$

$$= x^2 \gamma^2 \left(1 - \frac{u^2}{c^2} \right) - c^2 t^2 \gamma^2 \left(1 - \frac{u^2}{c^2} \right)$$

$$= x^2 - c^2 t^2 \tag{2.34}$$

where we noted that $\gamma^2 \left(1 - u^2/c^2 \right) = 1$ in the last step.

If we let

$$l = ict \tag{2.35}$$

then

$$l^2 = -c^2 t^2$$

and we can rewrite Eq. (2.31) as

$$x^2 + y^2 + z^2 + l^2 = x'^2 + y'^2 + z'^2 + l'^2 \tag{2.36}$$

where $l' = ict'$.

Eq. (2.36) implies that the four-dimensional vector $[x, y, z, l]$ is invariant under a Lorentz transformation. Because the length of this vector doesn't change, we can think of a Lorentz transformation as a "rotation" in this 4-D space.

Using Eq. (2.35) and letting

$$\beta = \frac{u}{c} \tag{2.37}$$

we can write the Lorentz transformation given by Eqs. (2.24) as

$$x_i' = x_i + (\gamma - 1)\alpha_i \alpha_j x_j + i\beta\gamma l\alpha_i$$

$$l' = \gamma \left(l - i\beta\alpha_j x_j \right) \tag{2.38}$$

Letting $x_4 = l$ and $x_4' = l'$, we can rewrite Eqs. (2.38) as

$$x_i' = x_i + (\gamma - 1)(\alpha_1 x_1 + \alpha_2 x_2 + \alpha_3 x_3)\alpha_i + i\beta\gamma\alpha_i x_4$$

$$x_4' = -i\beta\gamma(\alpha_1 x_1 + \alpha_2 x_2 + \alpha_3 x_3) + \gamma x_4 \tag{2.39}$$

We can write Eqs. (2.39) as the matrix equation

$$\begin{bmatrix} x_1' \\ x_2' \\ x_3' \\ x_4' \end{bmatrix} = \begin{bmatrix} 1 + (\gamma-1)\alpha_1\alpha_1 & (\gamma-1)\alpha_1\alpha_2 & (\gamma-1)\alpha_1\alpha_3 & i\beta\gamma\alpha_1 \\ (\gamma-1)\alpha_1\alpha_2 & 1 + (\gamma-1)\alpha_2\alpha_2 & (\gamma-1)\alpha_2\alpha_3 & i\beta\gamma\alpha_2 \\ (\gamma-1)\alpha_1\alpha_3 & (\gamma-1)\alpha_2\alpha_3 & 1 + (\gamma-1)\alpha_3\alpha_3 & i\beta\gamma\alpha_3 \\ -i\beta\gamma\alpha_1 & -i\beta\gamma\alpha_2 & -i\beta\gamma\alpha_3 & \gamma \end{bmatrix} \begin{bmatrix} x_1 \\ x_2 \\ x_3 \\ x_4 \end{bmatrix} \tag{2.40}$$

For the case of S' moving in the x-direction, we can set $\alpha_i = [1,0,0]$. Then Eq. (2.40) reduces to

$$\begin{bmatrix} x_1' \\ x_2' \\ x_3' \\ x_4' \end{bmatrix} = \begin{bmatrix} \gamma & 0 & 0 & i\beta\gamma \\ 0 & 1 & 0 & 0 \\ 0 & 0 & 1 & 0 \\ -i\beta\gamma & 0 & 0 & \gamma \end{bmatrix} \begin{bmatrix} x_1 \\ x_2 \\ x_3 \\ x_4 \end{bmatrix} \tag{2.41}$$

The matrix equations in Eqs. (2.40) and (2.41) are of the form

$$\mathbf{X}' = \mathbf{AX} \tag{2.42}$$

where $\mathbf{X}'$ and $\mathbf{X}$ are 4-vector column matrices and $\mathbf{A}$ is a 4×4 transformation matrix. It is easy to verify that the determinant of $\mathbf{A}$ is equal to 1 so that the length of the 4-vector $\mathbf{X}'$ is equal to the length of the 4-vector $\mathbf{X}$. Thus, the Lorentz transformation can be thought of as a rotation of the 4-vector $\mathbf{X}$.

The 4-vector **X** and the Lorentz transformation given by Eq. (2.40) are the source of the idea that we live in a 4-D space-time continuum. But the four dimensions of this space-time continuum are not of the same type, as can be seen in Eq. (2.40). It takes the two equations in Eqs. (2.39) to describe this matrix equation.

We can rewrite the vector Lorentz transformation in Eqs. (2.38) as

$$\mathbf{r}' = \mathbf{r} + (\gamma - 1)(\boldsymbol{\alpha} \bullet \mathbf{r})\boldsymbol{\alpha} + i\beta\gamma l\boldsymbol{\alpha}$$
$$l' = \gamma(l - i\beta\boldsymbol{\alpha} \bullet \mathbf{r}) \tag{2.43}$$

It is clear from Eq. (2.43) that the first three components of our 4-vector are really the components of a 3-D vector, while the fourth component is really a scalar proportional to time.

A quaternion is made up of a *scalar* part, l, and a *vector* part, $\mathbf{r}$. Therefore, it may be more natural to represent the Lorentz transformation using quaternions. We will explore this in the next section.

2.10 Quaternion Representation of the Lorentz Transformation

An introduction to quaternions is given in Appendix A. More details about quaternions can be found in one of my other books.[2]

The Lorentz transformation is given by Eqs. (2.43) as

$$\mathbf{r}' = \mathbf{r} + (\gamma - 1)(\boldsymbol{\alpha} \bullet \mathbf{r})\boldsymbol{\alpha} + i\beta\gamma l\boldsymbol{\alpha}$$
$$l' = \gamma(l - i\beta\boldsymbol{\alpha} \bullet \mathbf{r}) \tag{2.44}$$

where

$$\beta = \frac{u}{c} \tag{2.45}$$

$$\gamma = \frac{1}{\sqrt{1 - \beta^2}} \tag{2.46}$$

$$l = ict \tag{2.47}$$

---

[2] Richard E. Haskell, *Vector and Tensor Analysis By Example – Including Cartesian Tensors, Quaternions, and Matlab Examples*, ISBN: 978-1515153115, 2015.

Our goal is write the two Lorentz transformation equations in Eq. (2.44) as a single quaternion equation. Let the quaternions q and q' be given by

$$q = (l, \mathbf{r}) = l + \mathbf{r} \tag{2.48}$$

and

$$q' = (l', \mathbf{r}') = l' + \mathbf{r}' \tag{2.49}$$

Now let

$$Q = (s, \mathbf{w}) = s + \mathbf{w} \tag{2.50}$$

be a quaternion with scalar part s and vector part $\mathbf{w}$. We now ask the following question: What values of s and $\mathbf{w}$ will make the quaternion equation

$$q' = QqQ \tag{2.51}$$

represent the Lorentz transformation?

If we substitute Eqs. (2.50) and (2.48) in Eq. (2.51) and use the quaternion multiplication rules given in Appendix A, we can write

$$
\begin{aligned}
q' &= (s, \mathbf{w})(l, \mathbf{r})(s, \mathbf{w}) \\
&= (s, \mathbf{w})(ls - \mathbf{r} \cdot \mathbf{w}, \ l\mathbf{w} + s\mathbf{r} + \mathbf{r} \times \mathbf{w}) \\
&= (ls^2 - s\mathbf{r} \cdot \mathbf{w} - l\mathbf{w} \cdot \mathbf{w} - s\mathbf{w} \cdot \mathbf{r} - \mathbf{w} \cdot \mathbf{r} \times \mathbf{w}, \\
&\qquad + sl\mathbf{w} + s^2\mathbf{r} + s\mathbf{r} \times \mathbf{w} + ls\mathbf{w} - (\mathbf{r} \cdot \mathbf{w})\mathbf{w} \\
&\qquad + l\mathbf{w} \times \mathbf{w} + s\mathbf{w} \times \mathbf{r} + \mathbf{w} \times \mathbf{r} \times \mathbf{w}) \\
&= (l(s^2 - w^2) - 2s\mathbf{r} \cdot \mathbf{w}, \\
&\qquad + 2sl\mathbf{w} + s^2\mathbf{r} + s\mathbf{r} \times \mathbf{w} - (\mathbf{r} \cdot \mathbf{w})\mathbf{w} \\
&\qquad + s\mathbf{w} \times \mathbf{r} + w^2\mathbf{r} - (\mathbf{w} \cdot \mathbf{r})\mathbf{w}) \\
&= (l(s^2 - w^2) - 2s\mathbf{r} \cdot \mathbf{w}, \\
&\qquad + 2sl\mathbf{w} + (w^2 + s^2)\mathbf{r} - 2(\mathbf{r} \cdot \mathbf{w})\mathbf{w}) \\
&= (l', \mathbf{r}') \tag{2.52}
\end{aligned}
$$

from which

$$l' = (s^2 - w^2)l - 2s\mathbf{r} \cdot \mathbf{w} \tag{2.53}$$

and

$$\mathbf{r'} = (w^2 + s^2)\mathbf{r} - 2(\mathbf{r \cdot w})\mathbf{w} + 2s/\mathbf{w} \tag{2.54}$$

If we let $\mathbf{w} = w\mathbf{\alpha}$, we can write Eq. (2.50) as

$$Q = (s, w\mathbf{\alpha}) = s + w\mathbf{\alpha} \tag{2.55}$$

then Eqs. (2.53) and (2.54) can be written as

$$l' = (s^2 - w^2)l - 2sw\mathbf{r \cdot \alpha} \tag{2.56}$$

and

$$\mathbf{r'} - (w^2 + s^2)\mathbf{r} - 2w^2(\mathbf{r \cdot \alpha})\mathbf{\alpha} + 2sw/\mathbf{\alpha} \tag{2.57}$$

From Eqs. (2.44), we know that the Lorentz transformation is given by

$$l' = \gamma l - i\beta\gamma\mathbf{\alpha} \bullet \mathbf{r} \tag{2.58}$$

and

$$\mathbf{r'} = \mathbf{r} + (\gamma - 1)(\mathbf{\alpha} \bullet \mathbf{r})\mathbf{\alpha} + i\beta\gamma l\mathbf{\alpha} \tag{2.59}$$

Comparing Eqs. (2.56) and (2.58), we see that Eq. (2.56) will be the Lorentz transformation if we set

$$\gamma = s^2 - w^2 \tag{2.60}$$

and

$$2sw = i\beta\gamma \tag{2.61}$$

Also, by comparing Eqs. (2.57) and (2.59), we see that Eq. (2.57) will be the Lorentz transformation if we use Eq. (2.61) and set

$$w^2 + s^2 = 1 \tag{2.62}$$

and

$$1 - \gamma = 2w^2 \tag{2.63}$$

Note that if you substitute Eq. (2.60) into Eq. (2.63), you get Eq. (2.62). Thus, from Eq. (2.63), we find that

$$w = \sqrt{\frac{1 - \gamma}{2}} \tag{2.64}$$

and from Eqs. (2.62) and (2.64), we can write

$$s^2 = 1 - w^2 = 1 - \frac{1-\gamma}{2} = \frac{1+\gamma}{2}$$

from which

$$s = \sqrt{\frac{1+\gamma}{2}} \tag{2.65}$$

Substituting Eqs. (2.64) and (2.65) into Eq. (2.55), we obtain

$$Q = \frac{1}{\sqrt{2}}\left(\sqrt{1+\gamma} + a\sqrt{1-\gamma}\right) \tag{2.66}$$

Therefore, the Lorentz transformation is given by the quaternion equation

$$q' = QqQ \tag{2.67}$$

where Q is given by Eq. (2.66).

Note that Q in Eq. (2.66) is a unit quaternion (see Eq. (2.62). It can therefore be written as

$$Q = \cos\varphi + a\sin\varphi = e^{a\varphi} \tag{2.68}$$

where

$$\cos\varphi = s = \sqrt{\frac{1+\gamma}{2}} \tag{2.69}$$

and

$$\sin\varphi = w = \sqrt{\frac{1-\gamma}{2}} \tag{2.70}$$

Using the trigonometric identity $\sin 2\varphi = 2\sin\varphi\cos\varphi$ and Eq. (2.61), we can write

$$\sin 2\varphi = i\beta\gamma = \frac{i\beta}{\sqrt{1-\beta^2}} \tag{2.71}$$

It also follows that

$$\cos 2\varphi = \cos^2\varphi - \sin^2\varphi = \left(\frac{1+\gamma}{2}\right) - \left(\frac{1-\gamma}{2}\right) = \gamma \tag{2.72}$$

and, from Eqs. (2.71) and (2.72),

$$\tan 2\varphi = i\beta \tag{2.73}$$

from which

$$2\varphi = \tan^{-1}(i\beta) = \tan^{-1}\left(i\frac{u}{c}\right) \tag{2.74}$$

You may have wondered what motivated the choice of Eq. (2.51)

$$q' = QqQ \tag{2.75}$$

for the Lorentz transformation. In a different book[3], I show that quaternions are used to describe rotations in 3-D space. In particular, to rotate a vector **v** about a unit vector **u** through an angle θ you compute the quaternion

$$v' = (0, v') = q(0, v)q^* \tag{2.76}$$

where

$$q = \cos\frac{\theta}{2} + \sin\frac{\theta}{2}\mathbf{u} \tag{2.77}$$

and

$$q^* = \cos\frac{\theta}{2} - \sin\frac{\theta}{2}\mathbf{u} \tag{2.78}$$

is the conjugate of q.

By analogy, we see from Eqs. (2.68) and (2.74) that the Lorentz transformation given by Eq. (2.51) can be thought of as a rotation by the (imaginary) angle 2φ given in Eq. (2.74). This is the same 4-vector rotation described in Section 2.9. The rotation takes place in the $\mathbf{\alpha} - l$ plane.

We can find the Lorentz transformation that gives q in terms of q' by pre-multiplying and post-multiplying both sides of Eq. (2.75) by Q^{-1}. Thus,

---

[3] Richard E. Haskell, *Vectors and Tensors By Example – Including Cartesian Tensors, Quaternions, and Matlab Examples*, ISBN: 978-1515153115, 2015.

$$Q^{-1}q'Q^{-1} = Q^{-1}QqQQ^{-1} = q$$

or

$$q = Q^{-1}q'Q^{-1} \qquad (2.79)$$

For a unit quaternion (see Appendix A),

$$Q^{-1} = Q^* \qquad (2.80)$$

That is,

$$Q^{-1} = Q^* = \frac{1}{\sqrt{2}}\left(\sqrt{1+\gamma} - \boldsymbol{a}\sqrt{1-\gamma}\right) \qquad (2.81)$$

Chapter 3

Relativistic Dynamics

3.1 Relativistic Momentum

We have seen that the principle of relativity demands that the equations of physics have the same form in all inertial frames of reference. We must therefore re-examine the laws of dynamics in terms of the relativistic kinematics developed in the preceding chapter. For low velocities these laws must reduce to Newton's three laws of motion.

The first law essentially defines an inertial frame of reference. That is, it is one in which, in the absence of all external influence (i.e. forces), a body will remain in a state of rest or uniform motion in a straight line. The third law is a statement about forces, namely, that if body A acts on body B with a force $\mathbf{F}_{BA}$ then body B acts on body A with an equal and opposite force $\mathbf{F}_{AB} = -\mathbf{F}_{BA}$. Let us accept these laws as holding in the relativistic case.

Newton's second law of motion states that in an inertial frame of reference, as defined by the first law, the net force acting on a body is equal to the rate of change of momentum $\mathbf{p}$ where $\mathbf{p} = m\mathbf{v}$, the product of the mass times the velocity of the body. Now since this law involves quantities like velocity and time, which we have seen have special relativistic transformation laws, we must examine this second law with care.

Let us assume that for the relativistic case we can write the second law in the form $\mathbf{F} = d\mathbf{p}/dt$ and try to determine what $\mathbf{p}$ must be in order to be consistent with the Lorentz transformation.

Consider two isolated bodies A and B which are only interacting with each other. If $\mathbf{F}_{BA}$ is the force acting on A due to B and $\mathbf{F}_{BA}$ is the force acting on B due to A, then by Newton's third law and our assumption about the form of the second law we can write

$$\mathbf{F}_{AB} = \frac{d\mathbf{p}_A}{dt} = -\mathbf{F}_{BA} = -\frac{d\mathbf{p}_B}{dt}$$

or

$$\frac{d}{dt}(\mathbf{p}_A + \mathbf{p}_B) = 0$$

so that the total momentum of the two bodies is conserved. Since it must be conserved in all coordinate systems we can write

$$\frac{d}{dt}(\mathbf{p}_A + \mathbf{p}_B) = \frac{d}{dt}(\mathbf{p}_A' + \mathbf{p}_B') \qquad (3.1)$$

where the primed system is any inertial frame of reference.

Now let us suppose that body A is at rest (at least instantaneously) in the unprimed system and that the primed system is moving in the $+x_1$ direction with a speed u. That is, $\alpha_i = (1,0,0)$. Further let the velocity of body B be $(v_i)_B = (v_B,0,0)$ in the unprimed system and be $(v_i')_B = (v_B',0,0)$ in the primed system. From Eq. (2.27) these velocities are then related by the expression

$$v_B' = \frac{v_B - u}{1 - \dfrac{uv_B}{c^2}} \qquad (3.2)$$

Also, since $(v_i)_A = 0$ then $(v_i')_A = (v_A',0,0)$ where $v_A' = -u$.

We are looking for an expression for $\mathbf{p}$, which will be some function of the velocity of a body and will reduce to $m\mathbf{v}$ for low velocities. We will therefore assume that $\mathbf{p}$ is in the direction of $\mathbf{v}$ and so for the one dimensional case being considered $\mathbf{p} = (p,0,0)$ and Eq. (3.1) can be written as

$$\frac{d}{dt}[p(v_A) + p(v_B)] = \frac{d}{dt'}[p(v_A') + p(v_B')] \qquad (3.3)$$

Since $v_A = 0$ it follows that $p(v_A) = 0$ since for low velocities $p(v_A) = m_A v_A$. Also

$$\frac{dp(v_A')}{dt} = \frac{dp(v_A')}{dv_A'} \frac{dv_A'}{dt'} = 0$$

since $v_A' = -u = $ constant. Eq. (3.3) then reduces to

$$\frac{dp(v_R)}{dt} = \frac{dp(v_R')}{dt'}$$

or

$$\frac{dp(v_B)}{dv_B} \frac{dv_B}{dt} = \frac{dp(v_B')}{dv_B'} \frac{dv_B'}{dt} \frac{dt}{dt'} \qquad (3.4)$$

From Eq. (3.2)

$$\frac{dv_B'}{dt} = \frac{\left(1 - \dfrac{uv_B}{c^2}\right)\dfrac{dv_B}{dt} + \left(v_B - u\right)\dfrac{u}{c^2}\dfrac{dv_B}{dt}}{\left(1 - \dfrac{uv_B}{c^2}\right)^2}$$

or

$$\frac{dv_B'}{dt} = \frac{1}{\gamma^2 \left(1 - \dfrac{uv_B}{c^2}\right)^2} \frac{dv_B}{dt} \qquad (3.5)$$

where $\gamma = 1/\sqrt{1 - u^2/c^2}$. Since $t' = \gamma\left(t - \dfrac{u}{c^2}x_B\right)$ then

$$\frac{dt'}{dt} = \gamma\left(1 - \frac{uv_B}{c^2}\right) \qquad (3.6)$$

Substituting Eqs. (3.5) and (3.6) into (3.4), we obtain

$$\frac{dp(v_B)}{dv_B} = \frac{1}{\gamma^3 \left(1-\dfrac{uv_B}{c^2}\right)^3} \frac{dp(v_B')}{dv_B'}$$

(3.7)

Since Eq. (3.7) must hold for any v_B, let $v_B = v = u$ so that $v_B' = 0$. Then Eq. (3.7) can be written as

$$\frac{dp(v_B)}{dv_B} = \frac{K_0}{\left(1-\dfrac{v^2}{c^2}\right)^{3/2}}$$

(3.8)

where

$$K_0 = \left. \frac{dp(v_B')}{dv_B'} \right|_{v_B'=0}$$

Integrating Eq. (3.8), we obtain

$$p(v) = K_0 \frac{v}{\sqrt{1-v^2/c^2}} + const$$

Since $p(0) = 0$ and $p(v) = mv$ for $v/c << 1$ the constant of integration is zero and $K_0 = m$. Therefore, the relativistic expression for momentum is

$$p(v) = \frac{mv}{\sqrt{1-v^2/c^2}}$$

(3.9)

For arbitrary directions of motion the relativistic momentum can be written in the vector form

$$p_i = \frac{mv_i}{\sqrt{1-v^2/c^2}}$$

(3.10)

where $v^2 = v_i v_i$.

The expression for the relativistic momentum is often written in the non-relativistic form

$$p_i = m_R v_i$$

where

$$m_R = \frac{m}{\sqrt{1 - v^2/c^2}}$$

is called the *relativistic mass* (and m the *rest mass*) and increases with velocity. This introduction of a relativistic mass that increases with velocity in order to force the expression for momentum to be $m_R v_i$ is sometimes useful but not essential. It will not be used in the following sections so the mass in which appears will always be the rest mass.

3.2 Relativistic Energy

In the previous section we found that the force acting on a particle is given by $F_i = dp_i/dt$ where the momentum p_i is given by

$$p_i = \frac{mv_i}{\sqrt{1 - v^2/c^2}} \tag{3.11}$$

The work dW done by the force F_i in moving the particle through a distance dx_i is, by definition

$$dW = F_i dx_i$$
$$= \frac{dp_i}{dt} dx_i$$
$$= dp_i \frac{dx_i}{dt}$$
$$= v_i dp_i \tag{3.12}$$

From Eq. (3.11)

$$dp_i = \frac{\left(1 - v^2/c^2\right)^{1/2} m\, dv_i + mv_i \frac{1}{2}\left(1 - v^2/c^2\right)^{-1/2} \frac{2v}{c^2} dv}{\left(1 - v^2/c^2\right)}$$

or

$$dp_i = \frac{m}{\left(1-v^2/c^2\right)^{1/2}}\, dv_i + \frac{mvv_i}{c^2\left(1-v^2/c^2\right)^{3/2}}\, dv \qquad (3.13)$$

Since $v^2 = v_i v_i$, $vdv = v_i dv_i$ and from Eq. (3.13) we can write

$$v_i dp_i = \frac{mvdv}{\left(1-v^2/c^2\right)^{3/2}}\left(1-\frac{v^2}{c^2}+\frac{v^2}{c^2}\right)$$

$$= \frac{mvdv}{\left(1-v^2/c^2\right)^{3/2}} = dW \qquad (3.14)$$

Let us define the kinetic energy T of the particle, as in classical mechanics, as the work done in bringing the particle from rest to a velocity v. Then from Eq. (3.14) we have that

$$T = \int_0^v dW$$

$$= \int_0^v \frac{mvdv}{\left(1-v^2/c^2\right)^{3/2}}$$

$$= \frac{mc^2}{\sqrt{1-v^2/c^2}}\Bigg|_0^v$$

$$T = mc^2\left(\frac{1}{\sqrt{1-v^2/c^2}}-1\right) \qquad (3.15)$$

Note that for $v/c \ll 1$ this equation reduces to

$$T = mc^2\left(1+\frac{v^2}{2c^2}+\frac{3v^4}{8c^4}+...-1\right)$$

$$T = \frac{1}{2}mv^2$$

as in classical mechanics.

Now the conservation of momentum discussed in Section 3.1 has another important consequence. Consider the inelastic collision shown in Fig. 3.1. In the unprimed system, two equal masses approach each other with velocities v and $-v$. They collide and stick together so that after the collision the total mass M is at rest. (This is required because of the symmetry of the situation). The statement of the conservation of momentum for this case is thus

$$\frac{mv}{\sqrt{1-v^2/c^2}} - \frac{mv}{\sqrt{1-v^2/c^2}} = 0$$

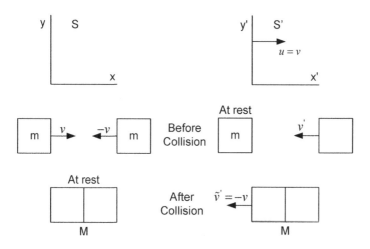

Fig. 3.1 Geometry of an inelastic collision

Now let us view this same inelastic collision from a reference frame moving to the right with a velocity $u = v$. Before the collision one mass is at rest in this frame while the other has a velocity

$$v' = \frac{-v-u}{1+\dfrac{uv}{c^2}} = -\frac{2v}{1+\dfrac{v^2}{c^2}} \tag{3.16}$$

After the collision the resultant mass M will be moving with a velocity $\tilde{v}' = -u = -v$.   The statement of the conservation of momentum in this frame is thus

$$\frac{mv'}{\sqrt{1-v'^2/c^2}} = \frac{M\tilde{v}'}{\sqrt{1-\tilde{v}'^2/c^2}}$$

or

$$\frac{-m2v}{\left(1+\dfrac{v^2}{c^2}\right)\sqrt{1-\dfrac{4v^2/c^2}{\left(1+\dfrac{v^2}{c^2}\right)^2}}} = \frac{-Mv}{\sqrt{1-v^2/c^2}}$$

or

$$\frac{2m}{\sqrt{\left(1-v^2/c^2\right)^2}} = \frac{M}{\sqrt{1-v^2/c^2}}$$

from which

$$M = \frac{2m}{\sqrt{\left(1-v^2/c^2\right)}} \tag{3.17}$$

We therefore see that the total mass of the system after the collision M is greater than the total mass of the system before the collision $2m$. We also note that the system loses kinetic energy as a result of the collision. This suggests that mass is a form of energy and that the loss in kinetic energy accounts for the increase in mass. If this is the case, we can define the total energy of a particle E to be equal to the sum of the kinetic energy plus the mass energy. The statement of the conservation of energy would then read (*kinetic energy* + *mass energy*) before collision = (*kinetic energy* + *mass energy*) after collision. We expect the mass energy to be some function of the mass $f(m)$.   Using Eqs. (3.15) and (3.17) and applying the conservation of energy to the unprimed frame we can write

$$2mc^2 \left(\frac{1}{\sqrt{1-v^2/c^2}} - 1 \right) + f(2m) = f(M)$$

or

$$\frac{2mc^2}{\sqrt{1-v^2/c^2}} - 2mc^2 = f\left(\frac{2m}{\sqrt{(1-v^2/c^2)}} \right) - f(2m)$$

We note that this equation will be satisfied if the mass energy $f(m) = mc^2$. Thus the total energy of a particle will be

$$E = T + mc^2$$

$$= mc^2 \left(\frac{1}{\sqrt{1-v^2/c^2}} - 1 \right) + mc^2$$

or

$$E = \frac{mc^2}{\sqrt{1-v^2/c^2}} \qquad (3.18)$$

There are several important relations between the momentum of a particle given by Eq. (3.11) and the energy of a particle given by Eq. (3.18). Since

$$p^2 = \frac{m^2 v^2}{1-v^2/c^2} \quad \text{and} \quad E^2 = \frac{m^2 c^4}{1-v^2/c^2}$$

it follows that

$$E^2 = \frac{p^2 c^4}{v^2} \quad \text{or} \quad \frac{p^2}{E^2} = \frac{v^2}{c^4}$$

from which

$$\frac{p_i}{E} = \frac{v_i}{c^2} \qquad (3.19)$$

Also note that

$$p^2 c^2 = \frac{m^2 v^2 c^4}{c^2 - v^2}$$

so that

$$p^2c^2 + m^2c^4 = m^2c^4 \left(\frac{c^2}{c^2 - v^2} \right) = \frac{m^2c^4}{1 - \frac{v^2}{c^2}} = E^2$$

or

$$E^2 = p^2c^2 + m^2c^4 \tag{3.20}$$

From Eq. (3.20), we can write

$$2E \frac{dE}{dt} = 2p_i c^2 \frac{dp_i}{dt}$$

or

$$\frac{dE}{dt} = \frac{p_i}{E} c^2 \frac{dp_i}{dt}$$

Using Eq. (3.19) and the relation $F_i = dp_i/dt$, we obtain

$$\frac{dE}{dt} = v_i F_i \tag{3.21}$$

3.3 Transformation of Energy and Momentum

The expressions (see Eqs. (3.11) and (3.18))

$$p_i = \frac{mv_i}{\sqrt{1 - v^2/c^2}} \tag{3.22}$$

and

$$E = \frac{mc^2}{\sqrt{1 - v^2/c^2}} \tag{3.23}$$

are the momentum and energy of a particle as measured in a certain reference frame (namely, one in which the velocity of the particle is v_i). An observer in a different reference frame will measure a different momentum and energy for this particle since he or she will measure a different velocity. In this new primed reference frame the momentum and energy are

$$p_i' = \frac{mv_i'}{\sqrt{1-v'^2/c^2}} \tag{3.24}$$

and

$$E' = \frac{mc^2}{\sqrt{1-v'^2/c^2}} \tag{3.25}$$

where v_i' is related to v_i by Eq. (2.25). In order to relate E' and p_i' directly to E and p_i we first need an expression for

$$\sqrt{1-v'^2/c^2}$$

in terms of the unprimed quantities. If we let $\alpha_j v_j = v_\parallel$ then from Eq. (2.25) we can write

$$v'^2 = v_i' v_i'$$

$$= \frac{\left\{\left[\delta_{ij} + (\gamma-1)\alpha_i\alpha_j\right]v_j - \gamma u\alpha_i\right\}\left\{\left[\delta_{ik} + (\gamma-1)\alpha_i\alpha_k\right]v_k - \gamma u\alpha_i\right\}}{\gamma^2\left(1 - \frac{u}{c^2}v_\parallel\right)^2}$$

$$= \frac{\left[\delta_{jk} + 2(\gamma-1)\alpha_j\alpha_k + (\gamma-1)^2\alpha_j\alpha_k\right]v_j v_k + \gamma^2 u^2 - 2\gamma^2 u\alpha_j v_j}{\gamma^2\left(1 - \frac{u}{c^2}v_\parallel\right)^2}$$

$$= \frac{v^2 + (\gamma^2-1)v_\parallel^2 + \gamma^2 u^2 - 2\gamma^2 uv_\parallel}{\gamma^2\left(1 - \frac{u}{c^2}v_\parallel\right)^2}$$

$$= \frac{\left(1 - \dfrac{u^2}{c^2}\right)v^2 + \dfrac{u^2}{c^2}v_\parallel^2 + u^2 - 2uv_\parallel}{\left(1 - \dfrac{uv_\parallel}{c^2}\right)^2}$$

Therefore,

$$1 - \frac{v'^2}{c^2} = \frac{c^2\left(1 - \dfrac{2uv_\parallel}{c^2} + \dfrac{u^2 v_\parallel^2}{c^4}\right) - \left(1 - \dfrac{u^2}{c^2}\right)v^2 - \dfrac{u^2}{c^2}v_\parallel^2 - u^2 + 2uv_\parallel}{c^2\left(1 - \dfrac{uv_\parallel}{c^2}\right)^2}$$

$$= \frac{c^2\left(1 - \dfrac{u^2}{c^2}\right) - \left(1 - \dfrac{u^2}{c^2}\right)v^2}{c^2\left(1 - \dfrac{uv_\parallel}{c^2}\right)^2}$$

$$= \frac{\left(1 - \dfrac{u^2}{c^2}\right)\left(1 - \dfrac{v^2}{c^2}\right)}{\left(1 - \dfrac{uv_\parallel}{c^2}\right)^2}$$

so that

$$\sqrt{1 - v'^2/c^2} = \frac{\sqrt{1 - \dfrac{u^2}{c^2}}\sqrt{1 - \dfrac{v^2}{c^2}}}{\left(1 - \dfrac{uv_\parallel}{c^2}\right)} \tag{3.26}$$

Using Eq. (3.26), we can write the energy E' given by Eq. (3.25) as

$$E' = \frac{\left(1 - \dfrac{uv_\parallel}{c^2}\right)mc^2}{\sqrt{1 - \dfrac{u^2}{c^2}}\sqrt{1 - \dfrac{v^2}{c^2}}}$$

or

$$E' = \frac{mc^2}{\sqrt{1-\frac{u^2}{c^2}}\sqrt{1-\frac{v^2}{c^2}}} - \frac{umv_{\parallel}}{\sqrt{1-\frac{u^2}{c^2}}\sqrt{1-\frac{v^2}{c^2}}}$$

$$= \frac{E}{\sqrt{1-\frac{u^2}{c^2}}} - \frac{u\alpha_j p_j}{\sqrt{1-\frac{u^2}{c^2}}}$$

The energy therefore transforms according to the relation

$$E' = \gamma\left(E - u\alpha_j p_j\right) \qquad (3.27)$$

or its inverse relation

$$E = \gamma\left(E' + u\alpha_j p_j'\right) \qquad (3.28)$$

Using Eqs. (2.25) and (3.26), we can write the momentum p_i' as

$$p_i' = \frac{mv_i'}{\sqrt{1-v'^2/c^2}}$$

$$= \frac{\left(1-\frac{uv_{\parallel}}{c^2}\right)m}{\sqrt{1-\frac{u^2}{c^2}}\sqrt{1-\frac{v^2}{c^2}}}\left[\frac{\left[\delta_{ij}+(\gamma-1)\alpha_i\alpha_j\right]v_j - \gamma u\alpha_i}{\gamma\left(1-\frac{uv_{\parallel}}{c^2}\right)}\right]$$

$$= \frac{\left[\delta_{ij}+(\gamma-1)\alpha_i\alpha_j\right]mv_j}{\sqrt{1-\frac{v^2}{c^2}}} - \frac{\gamma u\alpha_i mc^2}{c^2\sqrt{1-\frac{v^2}{c^2}}}$$

or,

$$p_i' = \left[\delta_{ij}+(\gamma-1)\alpha_i\alpha_j\right]p_j - \frac{\gamma u}{c^2}\alpha_i E \qquad (3.29)$$

with its inverse relation

$$p_i = \left[\delta_{ij} + (\gamma - 1)\alpha_i\alpha_j \right] p_j{}' + \frac{\gamma u}{c^2}\alpha_i E' \qquad (3.30)$$

Eqs. (3.27) - (3.30) are the transformation relations for momentum and energy.

3.4 Transformation Law for Force

The equation of motion of a particle in the unprimed frame is

$$F_i = \frac{dp_i}{dt} \qquad (3.31)$$

and in the primed frame is

$$F_i{}' = \frac{dp_i{}'}{dt} \qquad (3.32)$$

We wish to relate F_i and $F_i{}'$ directly. From Eq. (3.29) we can write

$$dp_i{}' = \left[\delta_{ij} + (\gamma - 1)\alpha_i\alpha_j \right] dp_j - \frac{\gamma u}{c^2}\alpha_i dE \qquad (3.33)$$

and from Eq. (2.24)

$$dt' = \gamma \left(dt - \frac{u}{c^2}\alpha_k dx_k \right) \qquad (3.34)$$

Thus,

$$F_i{}' = \frac{dp_i{}'}{dt'} = \frac{\left[\delta_{ij} + (\gamma - 1)\alpha_i\alpha_j \right] F_j - \frac{\gamma u}{c^2}\alpha_i \dfrac{dE}{dt}}{\gamma \left(1 - \dfrac{u}{c^2}\alpha_k v_k \right)}$$

or, using Eq. (3.21),

$$F_i' = \frac{\left[\delta_{ij} + (\gamma-1)\alpha_i\alpha_j\right]F_j - \dfrac{\gamma u}{c^2}\alpha_i v_j F_j}{\gamma\left(1 - \dfrac{u}{c^2}\alpha_k v_k\right)}$$

or

$$F_i' = \frac{1}{\gamma\left(1 - \dfrac{uv_\parallel}{c^2}\right)}\left\{F_i + \left[(\gamma-1)\alpha_i\alpha_j - \dfrac{\gamma u\alpha_i v_j}{c^2}\right]F_j\right\} \qquad (3.35)$$

Eq. (3.35) gives the force measured in the primed system in terms of the force in the unprimed system and the velocity of the particle in the *unprimed* system. We could obtain the inverse relation by interchanging primed and unprimed quantities and changing the sign of *u*. That would give us the force in the unprimed system in terms of the force in the primed system and the velocity of the particle in the *primed* system. However, it turns out to be convenient to have an expression for the force in the unprimed system in terms of the force in the primed system and the velocity of the particle in the *unprimed* system. This expression can be obtained simply by solving Eq. (3.35) for the unprimed force F_i. Thus, from Eq. (3.35), we can write

$$F_i = \gamma\left(1 - \frac{uv_\parallel}{c^2}\right)F_i' - \left[(\gamma-1)\alpha_i\alpha_j - \frac{\gamma u}{c^2}\alpha_i v_j\right]F_j \qquad (3.36)$$

Multiplying Eq. (3.36) through by α_i we obtain

$$\alpha_i F_i = \gamma\left(1 - \frac{uv_\parallel}{c^2}\right)\alpha_i F_i' - (\gamma-1)\alpha_j F_j + \frac{\gamma u}{c^2}v_j F_j$$

or

$$\alpha_j F_j = \left(1 - \frac{uv_\parallel}{c^2}\right)\alpha_j F_j' + \frac{u}{c^2}v_j F_j \qquad (3.37)$$

Substituting Eq. (3.37) into Eq. (3.36) we obtain

$$F_i = \gamma\left(1 - \frac{uv_\parallel}{c^2}\right)F_i' - (\gamma-1)\left[\left(1 - \frac{uv_\parallel}{c^2}\right)\alpha_i\alpha_jF_j' + \frac{u}{c^2}\alpha_iv_jF_j\right] + \frac{\gamma u}{c^2}\alpha_iv_jF_j$$

$$F_i = \gamma\left(1 - \frac{uv_\parallel}{c^2}\right)F_i' - (\gamma-1)\left(1 - \frac{uv_\parallel}{c^2}\right)\alpha_i\alpha_jF_j' + \frac{u}{c^2}\alpha_iv_jF_j \quad (3.38)$$

Multiplying Eq. (3.38) through by v_i we obtain

$$v_iF_i = \left(1 - \frac{uv_\parallel}{c^2}\right)\left[\gamma v_iF_i' - (\gamma-1)\alpha_i\alpha_jv_iF_j'\right] + \frac{u}{c^2}\alpha_iv_iv_jF_j$$

$$v_jF_j\left(1 - \frac{uv_\parallel}{c^2}\right) = \left(1 - \frac{uv_\parallel}{c^2}\right)\left[\gamma v_jF_j' - (\gamma-1)\alpha_kv_k\alpha_jF_j'\right]$$

$$v_jF_j = \gamma v_jF_j' - (\gamma-1)v_\parallel\alpha_jF_j' \quad (3.39)$$

Substituting Eq. (3.39) into Eq. (3.38), we obtain

$$F_i = \gamma\left(1 - \frac{uv_\parallel}{c^2}\right)F_i' - (\gamma-1)\left(1 - \frac{uv_\parallel}{c^2}\right)\alpha_i\alpha_jF_j'$$

$$+ \frac{u}{c^2}\gamma\alpha_iv_jF_j' - \frac{u}{c^2}(\gamma-1)v_\parallel\alpha_i\alpha_jF_j'$$

$$F_i = \gamma F_i' - (\gamma-1)\alpha_i\alpha_jF_j' - \frac{\gamma u}{c^2}\alpha_jv_jF_i' + \frac{\gamma u}{c^2}\alpha_iv_jF_j'$$

from which,

$$F_i = \left[\gamma\delta_{ij} - (\gamma-1)\alpha_i\alpha_j\right]F_j' + \frac{\gamma u}{c^2}\left(\alpha_iF_j' - \alpha_jF_i'\right)v_j \quad (3.40)$$

This is the equation we sought giving the force in the unprimed system F_i in terms of the force in the primed system F_i' and the velocity in the unprimed system v_j. This equation will be used to derive the Lorentz force for charged particles in Chapter 4.

Chapter 4

Maxwell's Equations

4.1 Coulomb's Law

Consider a charge q_1 located at $\mathbf{r}_1$ and a second charge q_2 located at $\mathbf{r}_2$. If $\hat{\mathbf{r}}_{12}$ is a unit vector in the direction of $\mathbf{r}_2 - \mathbf{r}_1$ and $r_{12} = |\mathbf{r}_2 - \mathbf{r}_1|$ then Coulomb's law states that the force on q_2 due to q_1 is given by

$$\mathbf{F}_e = k_e \frac{q_1 q_2}{r_{12}^2} \hat{\mathbf{r}}_{12} \qquad (4.1)$$

where the constant k_e depends on the units used. Electromagnetic unit can be confusing, and we will consider them in the next section. However, there are a few things to note about Eq. (4.1). Charles Coulomb discovered this law experimentally in 1784 using a torsion balance that he had designed. By Eq. (4.1), charges of the same sign repel each other and charges of unlike sign attract each other. There will obviously be a problem when $r_{12} = |\mathbf{r}_2 - \mathbf{r}_1| = 0$, because the force will become infinite. Eq. (4.1) essentially assumes that q_1 and q_2 are point charges, but the exact nature of charge is poorly understood. We talk of an electron having a certain amount of negative charge, and a proton having the same amount of positive charge. However, no one has ever seen an electron and exactly what an electron is remains a mystery. We will return to these thoughts about the nature of physical reality in Chapter 6, but for now we will return to the usual approach taken in physics: assume some model to represent physical reality (point charges), find some equations (Coulomb's law given by Eq. (4.1)) as see if the predictions agree with experiment.

4.2 Units

The most common system of units, taught in almost all engineering and science courses, is the SI[4] system, also called MKS, based on the three fundamental units, meter (m), kilograms (kg), and seconds (s). The advantage of the SI system is that it leads to practical values of units, which can be measured in the laboratory.

An older system, called *Gaussian* or CGS, uses centimeters (cm), grams (g), and seconds (s) for the three fundamental units. When limited to mechanics, the differences between the CGS and MKS are not significant, and conversion between the two is straightforward:

$$1\,\text{m} = 10^2\,\text{cm}$$
$$1\,\text{kg} = 10^3\,\text{g}$$

It is possible to define other units in terms of the three basic units of length (L), mass (M), and time (T). For example, from Newton's second law,

$$\mathbf{F} = m\mathbf{a} = m\frac{d\mathbf{v}}{dt}$$

we see that force has the units MLT^{-2}. In the SI system, this unit of force is $\text{kg}\cdot\text{m}\cdot\text{s}^{-2}$, which is called a *newton*, N. In the CGS system, this unit of force is $\text{g}\cdot\text{cm}\cdot\text{s}^{-2}$, which is called a *dyne*. Note that one dyne is equal to 10^{-5} newtons. Potential energy is force times distance and therefore has the units ML^2T^{-2}. Note that this is the same units as kinetic energy, mass time velocity squared. In the SI system, this unit of energy is $\text{kg}\cdot\text{m}^2\cdot\text{s}^{-2}$, or $\text{N}\cdot\text{m}$, which is called a *joule*, J. In the CGS system, this unit of energy is $\text{g}\cdot\text{cm}^2\cdot\text{s}^{-2}$, or $\text{dyne}\cdot\text{cm}$, which is called an *erg*. Note that one erg is equal to 10^{-7} joules.

However, when it comes to electrical units, the differences between the two systems of units become more significant. While it is possible to define the electrical units in terms of the three basic units of length (L), mass (M), and time (T), the SI system actually

[4] from the French Le Système International d'Unités.

defines the unit of current, the *ampere*, A, to be a fundamental unit.

The *ampere* is defined to be that current which, if flowing in two parallel conductors of infinite length one meter apart would produce a force of 2×10^{-7} newtons (N) on each meter of length. The conductors must also be of negligible circular cross section and the whole thing must be in a vacuum.

In the SI system, the unit of charge is the *coulomb*, C, and is defined to be the amount of charge moving through a wire in one second by a current of one ampere. Separating charges produces potential energy measured in joules. The electric potential is this potential energy per unit charge, measured in volts. One *volt* is one joule per coulomb.

In the SI system, the value of the constant k_e in Eq. (4.1) is given by

$$k_e = 1/4\pi\varepsilon_0 \qquad\qquad (4.2)$$

where $\varepsilon_0 = 8.85 \times 10^{-12} \, C^2/m^2 N$ is the *permittivity of free space*. Coulomb's law can then be written as

$$\mathbf{F}_e = \frac{q_1 q_2}{4\pi\varepsilon_0} \frac{\hat{\mathbf{r}}_{12}}{r_{12}^2} \quad (\text{SI}) \qquad\qquad (4.3)$$

Note that in the SI system of units, the constant ε_0 must have units to make the units of force in Eq. (4.3) come out right. The factor of 4π is included in Eq. (4.3) to keep it out of future equations.

In the CGS system, electrical units are not as straightforward as in the SI system. First of all, there are two electrical CGS subsystems: the electrostatic units, e.s.u., and the electromagnetic units, e.m.u.

In the CGS_{esu} subsystem, an electrical unit for current (or charge) is not defined to be one of the fundamental units. Rather, k_e in Eq. (4.1) is set equal to 1. This defines the CGS_{esu} *electrostatic* units. Coulomb's law can then be written in CGS_{esu} units as

$$\mathbf{F}_e = \frac{q_1 q_2}{r_{12}^2} \hat{\mathbf{r}}_{12} \quad (\text{CGS}_{esu}) \qquad\qquad (4.4)$$

A modified version of the *Gaussian* system of units, called the *rationalized Heaviside-Lorentz* system of units includes the factor 4π in Coulomb's law, Eq. (4.4). That is,

$$\mathbf{F}_{hl} = \frac{q_1 q_2}{4\pi r_{12}^2} \hat{\mathbf{r}}_{12} \quad \text{(HL)} \tag{4.5}$$

This defines charge to have the units $\left(ML^3T^{-2}\right)^{1/2} = M^{1/2}L^{3/2}T^{-1}$.

Using CGS units, the electrostatic unit of charge is $g^{1/2}cm^{3/2}s^{-1}$, which is called a *statcoulomb*, or sometimes just *esu*. This electrostatic unit of charge is also called a *franklin* (Fr). The electrostatic unit of voltage is the *statvolt*, where one statvolt is equal to one erg per statcoulomb.

In the CGS_{emu} subsystem, Coulomb's law is not used to define charge. Rather, Ampere's force law is used to define the force of attraction (or repulsion) between two current-carrying parallel conductors of length, *l*, separated by the distance, *d*, given by

$$F_m = 2k_m \frac{I_1 I_2 l}{d} \tag{4.6}$$

where I_1 and I_2 are the currents in the two conductors. If the currents are in the same direction, the conductors attract each other, and if the currents are in the opposite direction, the conductors repel each other. Recall that in the SI system of units, Eq. (4.6) is used to define the *ampere* as the fundamental unit of current.

From Eq. (4.1), we know that the units of $k_e q_1 q_2$ are $[force][length]^2$ or ML^3T^{-2}. Recall that current is charge per unit time. Therefore, if we substitute $I = q/T$ in Eq. (4.6), we see that $k_m q_1 q_2$ will have the units $[force][time]^2$ or ML. It therefore follows that the units of k_e/k_m will be $[length/time]^2$ or velocity squared. We can write this as

$$\frac{k_e}{k_m} = c^2 \tag{4.7}$$

We will see that c is the speed of light.

We saw in Eq. (4.4) that in the CGS_{esu} subsystem, the value of k_e in Eq. (4.1) is set equal to 1. In the CGS_{emu} subsystem, the value of k_m in Eq. (4.6) is set equal to 1. In this case, Eq. (4.6) can be rewritten as

$$F_m = 2\frac{I_1 I_2 l}{d} = 2\frac{q_{1m} q_{2m} l}{T^2 d} \quad (CGS_{emu}) \tag{4.8}$$

Note that the electromagnetic (e.m.u.) unit of charge in Eq. (4.8) will have the units

$$Q_{emu} \equiv (ML)^{1/2} = M^{1/2}L^{1/2} \tag{4.9}$$

Recall from Eq. (4.4), that the electrostatic (e.s.u.) unit of charge will have the units

$$Q_{esu} \equiv (ML^3 T^{-2})^{1/2} = M^{1/2}L^{3/2}T^{-1} \tag{4.10}$$

From Eqs. (4.9) and (4.10), we see that the ratio of the electrostatic and electromagnetic units of charge will be

$$Q_{esu}/Q_{emu} \equiv LT^{-1}$$

which is the unit of velocity. That is,

$$Q_{esu}/Q_{emu} = c \tag{4.11}$$

Maxwell designed and built the device shown in Fig. 4.1 to measure the ratio of the electrostatic and electromagnetic units of charge given by Eq. (4.11). In the device, two parallel plates of a capacitor are charged so as to repel each other. Coils are attached to the capacitor plates such that the current through the coils cause attraction. When these two forces balanced each other, Maxwell measured the resistance in absolute units of $[length/time]$ and found the value to be very close to the known value of the speed of light, c.

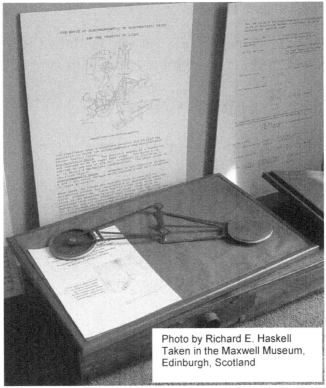

Photo by Richard E. Haskell
Taken in the Maxwell Museum,
Edinburgh, Scotland

Fig. 4.1 James Clerk Maxwell built this device to measure the ratio or the electrostatic and electromagnetic units of charge. The measured value was close to the speed of light.

This was a remarkable result. By a purely mechanical experiment involving static electric and magnetic fields, Maxwell measured the speed of light! This convinced him that light was an electromagnetic wave. His equations actually predicted this, but most scientists at the time were skeptical because no one knew how to generate and detect electromagnetic waves. It wasn't until 1888, nine years after Maxwell's death, that Heinrich Hertz generated and detected electromagnetic waves.

The *Gaussian* units are a combination of both the CGS_{esu} and CGS_{emu} subsystems. The unit of charge is the statcoulomb, or esu, or *franklin* (Fr). Using Eq. (4.10), we can write the unit of current, called a *statampere*, as

$$I_{esu} = Q_{esu}/T \equiv M^{1/2}L^{3/2}T^{-2} \qquad (4.12)$$

There is a corresponding current in the CGS_{emu} subsystem, called an *abampere*, which, from Eq. (4.9), is given by

$$I_{emu} = Q_{emu}/T = (ML)^{1/2}T^{-1} = M^{1/2}L^{1/2}T^{-1} \qquad (4.13)$$

Recall that in the SI system of units, the *ampere* is defined to be that current which, if flowing in two parallel conductors of infinite length one meter apart would produce a force of 2×10^{-7} newtons (N) on each meter of length. Using this definition in Eq. (4.6), and redefining the contant k_m to be

$$k_m = \mu_0/4\pi \qquad (4.14)$$

we can write

$$2\times10^{-7}\,\text{N} = 2\frac{\mu_0}{4\pi}\frac{1\text{A}1\text{A}1\text{m}}{1\text{m}}$$

from which

$$\mu_0 = 4\pi\times10^{-7}\,\text{N}/\text{A}^2$$
$$= 4\pi\times10^{-7}\,\text{kg}\cdot\text{m}\cdot\text{s}^{-2}\cdot\text{A}^{-2} \qquad (4.15)$$

is called the *permeability of free space*.

If we substitute Eqs. (4.2) and (4.14) in Eq. (4.7), we obtain

$$\frac{k_e}{k_m} = \frac{1}{\mu_0\varepsilon_0} = c^2 \qquad (4.16)$$

Using Eq. (4.15) and the known value of the speed of light, c, we can solve Eq. (4.16) for the value of the permittivity of free space ε_0 as

$$\varepsilon_0 = \frac{1}{\mu_0 c^2} = \frac{1}{4\pi \times 10^{-7} \, \text{N} \cdot \text{A}^{-2} \times (2.997925 \times 10^8)^2 \, \text{m}^2 \cdot \text{s}^{-2}}$$

or

$$\varepsilon_0 \cong 8.854 \times 10^{-12} \, \text{N}^{-1} \cdot \text{m}^{-2} \cdot \text{A}^2 \cdot \text{s}^2$$

or

$$\varepsilon_0 \cong 8.854 \times 10^{-12} \, \text{C}^2/\text{m}^2 \text{N} \qquad (4.17)$$

Note that the constants ε_0 and μ_0 exist only in the SI system of units to make the units work. These constants do not occur in the Gaussian system of units.

In the Gaussian system of units, $k_e = 1$, and from Eq. (4.7),

$$k_m = 1/c^2 \qquad (4.18)$$

How many *statamperes* are equivalent to 1 ampere? Substituting Eq. (4.18) in Eq. (4.6) and using the definition of an ampere, we can write

$$2 \times 10^{-7} \, \text{N} = 2 \frac{1}{c^2} \frac{I_{sA}^2 \, \text{1m}}{\text{1m}}$$

from which

$$I_{sA}^2 = 10^{-7} \, \text{N}c^2 = 10^{-7} \frac{\text{kg} \cdot \text{m}}{\text{s}^2} \times (3 \times 10^8)^2 \frac{\text{m}^2}{\text{s}^2} \times 10^3 \frac{\text{g}}{\text{kg}} \times 10^6 \frac{\text{cm}^3}{\text{m}^3}$$

or

$$I_{sA} = 3 \times 10^9 \text{ statamperes}$$

Therefore,

$$1 \text{ ampere} = 3 \times 10^9 \text{ statamperes} \qquad (4.19)$$

It follows that

$$1 \text{ coulomb} = 3 \times 10^9 \text{ statcoulomb} \qquad (4.20)$$

Recall that the electrostatic unit of voltage is the *statvolt*, where one statvolt is equal to one erg per statcoulomb, and in the SI units, one *volt* is one joule per coulomb. Using Eq. (4.20) and the fact that one erg is equal to 10^{-7} joules, is easy to show that one statvolt is equal to 300 volts.

How many *abamperes* are equivalent to 1 ampere? Using the value of $k_m = 1$ in Eq. (4.6) and using the definition of an ampere, we can write

$$2 \times 10^{-7} \, N = 2 \frac{I_{abA}^2 \, 1m}{1m}$$

From which it is easy to show that

$$1 \text{ ampere} - 0.1 \text{ abamperes} \qquad (4.21)$$

and

$$1 \text{ coulomb} = 0.1 \text{ abcoulomb} \qquad (4.22)$$

The emu unit of voltage is the *abvolt*, where one abvolt is equal to one erg per abcoulomb. It is easy to show that one abvolt is equal to 10^{-8} volts.

4.3 The Electric Field

If we define the electric field **E** due to a charge Q to be the force per unit charge on a test charge q, then from Eq. (4.5) we can write

$$\mathbf{E} = \frac{\mathbf{F}}{q} = \frac{Q}{4\pi r^2} \boldsymbol{\alpha} \qquad (4.23)$$

where we have used *rationalized Heaviside-Lorentz* units and r is the distance between Q and q, and $\boldsymbol{\alpha}$ is a unit vector in the direction of q from Q. Note that the unit of **E** is dynes per esu or dynes per statcoulomb, which is $g^{1/2} cm^{-1/2} s^{-1}$, also the same as statvolts per centimeter. The SI unit of **E** is volts per meter.

We can think of the force on q as being proportional to **E**, and **E** is proportional to Q and is inversely proportional to the square of r. If we consider a sphere to be centered at Q then we can picture the electric field vector as shown in Fig. 4.2, where the number of electric field lines is proportional to Q and the force on a test charge q will be proportional to the density of the electric field lines per unit area. This force must decrease as $1/r^2$ inasmuch as the area of a sphere increases as r^2 and the number of electric field lines is constant for a given value of Q.

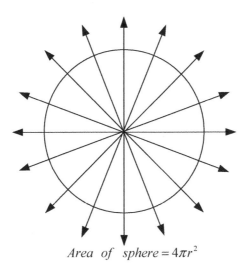

Area of sphere $= 4\pi r^2$

Fig. 4.2 Electric field of a point charge Q

The electric flux Ψ_E on a closed surface S is defined as

$$\Psi_E = \oiint_S \mathbf{E} \cdot \mathbf{n} \, dA \tag{4.24}$$

where $\mathbf{n}$ is an outward unit vector normal to the area dA. In Fig. 4.2, the unit vectors $\boldsymbol{\alpha}$ and $\mathbf{n}$ in Eqs. (4.23) and (4.24) will be parallel, and therefore, from Eq. (4.23), we can write

$$\mathbf{E} \cdot \mathbf{n} = \frac{Q}{4\pi r^2} \tag{4.25}$$

Therefore, the electric flux Ψ_E in Eq. (4.24), can be written as

$$\Psi_E = \oiint_S \mathbf{E} \cdot \mathbf{n} dA = \oiint_S \frac{Q}{4\pi r^2} dA = Q \qquad (4.26)$$

since the area of the sphere in Fig. 4.2 is $4\pi r^2$. Therefore, the total electric flux leaving the closed spherical surface in Fig. 4.2 is equal to the total charge within the closed surface. This is a general statement of *Gauss's Law* as we will see below.

More generally, we can consider a charge density ρ such that the charge Q within a small volume element ΔV is given by

$$Q = \rho \Delta V \qquad (4.27)$$

The *divergence* of the vector $\mathbf{E}$ is defined by[5]

$$\nabla \bullet \mathbf{E} = \lim_{\Delta V \to 0} \frac{\Psi_E}{\Delta V} = \lim_{\Delta V \to 0} \frac{1}{\Delta V} \oiint_S \mathbf{E} \cdot \mathbf{n} dA \qquad (4.28)$$

from which, using Eqs. (4.26) and (4.27), we can write

$$\nabla \bullet \mathbf{E} \Delta V = \rho \Delta V \qquad (4.29)$$

Taking the limit as $\Delta V \to 0$ and integrating over the volume enclosed by the closed surface S, we can write

$$\oiiint_V \nabla \bullet \mathbf{E} dV = \oiiint_V \rho dV \qquad (4.30)$$

or

$$\oiiint_V (\nabla \bullet \mathbf{E} - \rho) dV = 0 \qquad (4.31)$$

Since the volume V is arbitrary, the integrand in Eq. (4.31) must be zero. We therefore conclude that

$$\nabla \bullet \mathbf{E} = \rho \qquad (4.32)$$

---

[5] See, for example, Richard E. Haskell, *Vectors and Tensors By Example*, ISBN: 1515153115, 2015.

which is the first of Maxwell's equations.

The divergence theorem is given by[5]

$$\oint\!\!\!\oint_S \mathbf{E} \cdot \mathbf{n} dS = \oint\!\!\!\oint\!\!\!\oint_V \nabla \cdot \mathbf{E} dV \tag{4.33}$$

Substituting Eq. (4.30) into Eq. (4.33), we obtain

$$\Psi_E = \oint\!\!\!\oint_S \mathbf{E} \cdot \mathbf{n} dS = \oint\!\!\!\oint\!\!\!\oint_V \rho dV \tag{4.34}$$

which is the more general form of Gauss's Law.

Electric Potential

Let charge Q be at the origin and define the scalar potential

$$\varphi = \frac{Q}{4\pi r} \tag{4.35}$$

If we take the gradient of φ in spherical coordinates we can write

$$\nabla \varphi = \frac{Q}{4\pi} \boldsymbol{\alpha} \frac{d}{dr}\left(\frac{1}{r}\right) = -\frac{Q}{4\pi r^2} \boldsymbol{\alpha} \tag{4.36}$$

Comparing Eqs. (4.23) and (4.36) we see that we can write $\mathbf{E}$ as

$$\mathbf{E} = -\nabla \varphi \tag{4.37}$$

Because $\mathbf{E}$ can be written as the negative gradient of a scalar, it follows that

$$\nabla \times \mathbf{E} = 0 \tag{4.38}$$

since the curl of a gradient is always zero. You can verify this by writing the components of the curl as[6]

$$\varepsilon_{ijk} \frac{\partial E_k}{\partial x_j} = \varepsilon_{ijk} E_{k,j} \qquad (4.39)$$

and the components of the gradient as

$$\frac{\partial \varphi}{\partial x_k} = \varphi_{,k} \qquad (4.40)$$

Then writing the components of Eq. (4.37) as $E_k = -\varphi_{,k}$ we can write the curl of the gradient from Eq. (4.39) as

$$\varepsilon_{ijk} E_{k,j} = -\varepsilon_{ijk} \varphi_{,kj} = 0 \qquad (4.41)$$

which is easily seen to be zero by summing the repeated indices j and k from 1 to 3 and noting that $\varphi_{,kj} = \varphi_{,jk}$ for all j and k.

We therefore see from Coulomb's law that when the charges are at rest the electric field **E** satisfies the electrostatic equations (4.38) and (4.32), which can be written in component form as

$$\varepsilon_{ijk} \frac{\partial E_k}{\partial x_j} = 0 \qquad (4.42)$$

and

$$\frac{\partial E_i}{\partial x_i} = \rho \qquad (4.43)$$

---

[6] ε_{ijk} is the alternating unit tensor that is equal to +1 if i, j, k are in cyclic order, -1 if i, j, k are in noncyclic order, and 0 if any two subscripts are repeated. We also use the comma notation $\dfrac{\partial}{\partial x_j} = _{,j}$. For a complete discussion of Cartesian tensors, see, Richard E. Haskell, *Vectors and Tensors By Example*, ISBN: 1515153115, 2015.

4.4 The Lorentz Force

Let a collection of source charges with a charge density ρ' be at rest in a reference frame S' which has velocity $u_i = u\alpha_i$ and moves uniformly relative to another reference frame S. This charge density gives rise to an electrostatic field E_i' in the frame S' and will produce a force $F_i' = q'E_i'$ on a test charge q'. The force on a test charge q as measured by observers in reference frame S can be found by using the force transformation given by Eq. (3.40).

Two cases can be distinguished. If the velocity v_i of the test charge is zero, the electric field E_i measured in S is defined by the force relation $F_i = qE_i$. Assuming that charge is invariant (i.e., $q = q'$) and setting $F_i = qE_i$, $F_j' = q'E_j'$ and $v_j = 0$ in Eq. (3.40), we find that E_i is given by

$$E_i = \left[\gamma\delta_{ij} - (\gamma - 1)\alpha_i\alpha_j \right] E_j'$$

(4.44)

If the test charge q is now allowed to move with a velocity v_i then Eq. (3.40) becomes

$$F_i = \left[\gamma\delta_{ij} - (\gamma - 1)\alpha_i\alpha_j \right] qE_j' + \frac{q\gamma\beta}{c}\left(\alpha_i E_j' - \alpha_j E_i' \right) v_j$$

(4.45)

which, using Eq. (4.44), can be written as

$$F_i = qE_i + \frac{q}{c} C_{ij} v_j$$

(4.46)

where

$$C_{ij} = \gamma\beta\left(\alpha_i E_j' - \alpha_j E_i' \right)$$

(4.47)

Since C_{ij} is anti-symmetric, an axial vector B_i, called the *magnetic field*, can be defined by the relation[7]

$$C_{ij} = \varepsilon_{ijk}B_k \tag{4.48}$$

Therefore, Eq. (4.46) can be written as

$$F_i = q\left(E_i + \frac{1}{c}\varepsilon_{ijk}v_j B_k\right) \tag{4.49}$$

or

$$\mathbf{F} = q\left(\mathbf{E} + \frac{1}{c}\mathbf{v}\times\mathbf{B}\right) \tag{4.50}$$

which is the *Lorentz force*. Note that the magnetic field is directly related to the electric field in the moving frame of reference.

Note that in the *rationalized Heaviside-Lorentz* units we are using, the unit of the magnetic field, **B**, in Eq. (4.50) is the same as the unit of **E**, namely, $g^{1/2}cm^{-1/2}s^{-1}$. When referring to **B**, often called the *magnetic flux density*, this emu unit is called a *gauss*.

A variety of relationships exists between the field vectors E_i', E_i, and B_i. These relations will be used in Section 4.5 to aid in the derivation of Maxwell's equations. By multiplying Eq. (4.47) by ε_{rij} and using Eq. (4.48), we can write[8]

$$\varepsilon_{rij}\varepsilon_{ijk}B_k = \gamma\beta\left(\varepsilon_{rij}\alpha_i E_j' - \varepsilon_{rij}\alpha_j E_i'\right)$$

$$\varepsilon_{ijr}\varepsilon_{ijk}B_k = \gamma\beta\left(\varepsilon_{rij}\alpha_i E_j' + \varepsilon_{rji}\alpha_j E_i'\right)$$

---

[7] Note that $C_{ij} = \varepsilon_{ijk}B_k = \varepsilon_{ij1}B_1 + \varepsilon_{ij2}B_2 + \varepsilon_{ij3}B_3 = \begin{pmatrix} 0 & B_3 & -B_2 \\ -B_3 & 0 & B_1 \\ B_2 & -B_1 & 0 \end{pmatrix}$

[8] In step 3 we use the general relationship $\varepsilon_{ijk}\varepsilon_{irs} = \delta_{jr}\delta_{ks} - \delta_{js}\delta_{kr}$

$$\left(\delta_{jj}\delta_{rk} - \delta_{jk}\delta_{rj}\right)B_k = \gamma\beta\left(\varepsilon_{rij}\alpha_i E_j ' + \varepsilon_{rij}\alpha_i E_j '\right)$$

$$\left(3\delta_{rk} - \delta_{rk}\right)B_k = 2\gamma\beta\varepsilon_{rij}\alpha_i E_j '$$

$$B_r = \gamma\beta\varepsilon_{rij}\alpha_i E_j '$$

$$B_r = \gamma\frac{u}{c}\varepsilon_{rij}\alpha_i E_j ' \tag{4.51}$$

which is the same as

$$\mathbf{B} = \frac{\gamma}{c}\left(\mathbf{u}\times\mathbf{E}'\right) \tag{4.52}$$

Multiplying Eq. (4.44) by the unit vector α_i, leads to

$$\alpha_i E_i = \alpha_i E_i ' \tag{4.53}$$

and thus the component of $\mathbf{E}$ in the direction of motion is invariant. Also, multiplying Eq. (4.44) by $\varepsilon_{rsi}\alpha_s$ leads to[9]

$$\varepsilon_{rsi}\alpha_s E_i = \gamma\varepsilon_{rsj}\alpha_s E_j ' \tag{4.54}$$

and thus the component of $\mathbf{E}$ perpendicular to the direction of motion is larger than the corresponding component of $\mathbf{E}'$ by the factor γ.

From Eqs. (4.51) and (4.54) an expression relating B_i and E_i can be written as

$$B_i = \frac{u}{c}\varepsilon_{ijk}\alpha_j E_k = \beta\varepsilon_{ijk}\alpha_j E_k \tag{4.55}$$

Multiplying Eq. (4.47) by α_j and using Eqs. (4.48) and (4.53), we can write

---

[9] Note that $\varepsilon_{rsi}\alpha_s\alpha_i = 0$

$$\alpha_j C_{ij} = \gamma\beta\left(\alpha_j\alpha_i E_j{}' - \alpha_j\alpha_j E_i{}'\right)$$

$$\alpha_j\varepsilon_{ijk}B_k = \gamma\beta\left(\alpha_j\alpha_i E_j - \alpha_j\alpha_j E_i{}'\right)$$

from which

$$E_i{}' = \alpha_i\alpha_j E_j - \frac{1}{\gamma\beta}\varepsilon_{ijk}\alpha_j B_k \tag{4.56}$$

Using Eq. (4.53), Eq. (4.44) can also be written as

$$E_i{}' = \frac{1}{\gamma}E_i + \frac{(\gamma-1)}{\gamma}\alpha_i\alpha_j E_j \tag{4.57}$$

Equating the right-hand sides of Eqs. (4.56) and (4.57), we can write

$$\alpha_i\alpha_j E_j - \frac{1}{\gamma\beta}\varepsilon_{ijk}\alpha_j B_k = \frac{1}{\gamma}E_i + \frac{(\gamma-1)}{\gamma}\alpha_i\alpha_j E_j$$

$$-\frac{1}{\beta}\varepsilon_{ijk}\alpha_j B_k = E_i - \alpha_i\alpha_j E_j$$

$$\alpha_i\alpha_j E_j = E_i + \frac{1}{\beta}\varepsilon_{ijk}\alpha_j B_k \tag{4.58}$$

Substituting Eq. (4.58) into Eq. (4.56), and using $\beta^2 = \left(\gamma^2-1\right)/\gamma^2$ we obtain

$$E_i{}' = E_i + \frac{1}{\beta}\varepsilon_{ijk}\alpha_j B_k - \frac{1}{\gamma\beta}\varepsilon_{ijk}\alpha_j B_k$$

$$E_i{}' = E_i + \frac{1}{\beta}\left(\frac{\gamma-1}{\gamma}\right)\varepsilon_{ijk}\alpha_j B_k$$

$$E_i{}' = E_i + \frac{\gamma}{\gamma+1}\beta\varepsilon_{ijk}\alpha_j B_k \tag{4.59}$$

4.5 Maxwell's Equations

Since E_i' is a static field in the S' reference frame, it satisfies the electrostatic equations (see Eqs. (4.42) and (4.43))

$$\varepsilon_{ijk} \frac{\partial E_k'}{\partial x_j'} = 0 \tag{4.60}$$

and

$$\frac{\partial E_i'}{\partial x_i'} = \rho' \tag{4.61}$$

In this section, it will be shown that these two equations transform into the four Maxwell equations which describe the time and spatial variations of E_i and B_i in the S frame.

First it is necessary to relate time and spatial variations in the two frames of reference. Consider some function $f(x_i',t')$. From the Lorentz transformations given in Eq. (2.24) we see that both x_i' and t' will be functions of x and t. Using Eq. (2.24) we can write

$$\frac{\partial f}{\partial x_i} = \frac{\partial f}{\partial x_j'}\frac{\partial x_j'}{\partial x_i} + \frac{\partial f}{\partial t'}\frac{\partial t'}{\partial x_i}$$

$$= \left[\delta_{ij} + (\gamma-1)\alpha_i\alpha_j\right]\frac{\partial f}{\partial x_j'} - \frac{\gamma\beta}{c}\alpha_i\frac{\partial f}{\partial t'} \tag{4.62}$$

and

$$\frac{\partial f}{\partial t} = \frac{\partial f}{\partial t'}\frac{\partial t'}{\partial t} + \frac{\partial f}{\partial x_j'}\frac{\partial x_j'}{\partial t}$$

$$= \gamma\frac{\partial f}{\partial t'} - \gamma\beta c\alpha_j\frac{\partial f}{\partial x_j'} \tag{4.63}$$

In a similar way, we can write

$$\frac{\partial f}{\partial x_i'} = \frac{\partial f}{\partial x_j}\frac{\partial x_j}{\partial x_i'} + \frac{\partial f}{\partial t}\frac{\partial t}{\partial x_i'}$$

$$= \left[\delta_{ij} + (\gamma-1)\alpha_i\alpha_j\right]\frac{\partial f}{\partial x_j} + \frac{\gamma\beta}{c}\alpha_i\frac{\partial f}{\partial t} \tag{4.64}$$

and

$$\frac{\partial f}{\partial t'} = \frac{\partial f}{\partial t}\frac{\partial t}{\partial t'} + \frac{\partial f}{\partial x_j}\frac{\partial x_j}{\partial t'}$$

$$= \gamma\frac{\partial f}{\partial t} + \gamma\beta c\alpha_j\frac{\partial f}{\partial x_j} \qquad (4.65)$$

When f is a static field in S' [i.e., $f \neq f(t')$], then Eqs. (4.62) and (4.63) reduce to

$$\frac{\partial f}{\partial x_i} = \left[\delta_{ij} + (\gamma-1)\alpha_i\alpha_j\right]\frac{\partial f}{\partial x_j'} \qquad (4.66)$$

and

$$\alpha_j\frac{\partial f}{\partial x_j'} = -\frac{1}{\gamma\beta c}\frac{\partial f}{\partial t} \qquad (4.67)$$

and Eq. (4.65) reduces to

$$\frac{\partial f}{\partial t} = -\beta c\alpha_j\frac{\partial f}{\partial x_j} \qquad (4.68)$$

Substituting Eq. (4.68) into Eq. (4.64), we obtain

$$\frac{\partial f}{\partial x_i'} = \left[\delta_{ij} + (\gamma-1)\alpha_i\alpha_j\right]\frac{\partial f}{\partial x_j} - \gamma\beta^2\alpha_i\alpha_j\frac{\partial f}{\partial x_j}$$

$$\frac{\partial f}{\partial x_i'} = \left[\delta_{ij} - \frac{(\gamma-1)}{\gamma}\alpha_i\alpha_j\right]\frac{\partial f}{\partial x_j} \qquad (4.69)$$

where $\beta^2 = (\gamma^2-1)/\gamma^2$ was used in the last step.

Consider first B_i and form the divergence of B_i using Eqs. (4.66) and (4.51). Thus

$$\frac{\partial B_i}{\partial x_i} = \left[\delta_{ir} + (\gamma-1)\alpha_i\alpha_r\right]\gamma\beta\varepsilon_{ijk}\alpha_j\frac{\partial E_k'}{\partial x_r'}$$

$$\frac{\partial B_i}{\partial x_i} = -\gamma\beta\alpha_j\varepsilon_{jrk}\frac{\partial E_k{}'}{\partial x_r{}'} + \gamma\beta(\gamma-1)\varepsilon_{ijk}\alpha_j\alpha_i\alpha_r\frac{\partial E_k{}'}{\partial x_r{}'}$$

$$\frac{\partial B_i}{\partial x_i} = 0 \qquad\qquad (4.70)$$

since the first term is zero by Eq. (4.60) and in the second term $\varepsilon_{ijk}\alpha_j\alpha_i = 0$.

Therefore, the divergence of B_i is zero in the S frame because the curl of $E_i{}'$ is zero in the S' frame.

Now form the curl of $E_i{}'$ as in Eq. (4.60) and use Eqs. (4.69) and (4.68) to write

$$\varepsilon_{ijk}\frac{\partial E_k{}'}{\partial x_j{}'} = 0 = \varepsilon_{ijk}\left[\delta_{js} - \frac{(\gamma-1)}{\gamma}\alpha_j\alpha_s\right]\frac{\partial E_k{}'}{\partial x_s}$$

$$= \varepsilon_{ijk}\left[\frac{\partial E_k{}'}{\partial x_j} + \frac{(\gamma-1)}{\gamma\beta c}\alpha_j\frac{\partial E_k{}'}{\partial t}\right] \qquad (4.71)$$

Substituting Eq. (4.59) for $E_k{}'$ in the first term of Eq. (4.71) and using Eq. (4.51) for B_i in the second term of Eq. (4.71), we can write

$$\varepsilon_{ijk}\frac{\partial E_k}{\partial x_j} + \frac{\gamma}{\gamma+1}\beta\varepsilon_{ijk}\varepsilon_{krs}\alpha_r\frac{\partial B_s}{\partial x_j} + \frac{(\gamma-1)}{c\gamma^2\beta^2}\frac{\partial B_i}{\partial t} = 0$$

$$\varepsilon_{ijk}\frac{\partial E_k}{\partial x_j} + \frac{\gamma}{\gamma+1}\beta\left(\delta_{ir}\delta_{js} - \delta_{is}\delta_{jr}\right)\alpha_r\frac{\partial B_s}{\partial x_j} + \frac{(\gamma-1)}{c\gamma^2\beta^2}\frac{\partial B_i}{\partial t} = 0$$

$$\varepsilon_{ijk}\frac{\partial E_k}{\partial x_j} + \frac{\gamma}{\gamma+1}\beta\left(\alpha_i\frac{\partial B_j}{\partial x_j} - \alpha_j\frac{\partial B_i}{\partial x_j}\right) + \frac{1}{c(\gamma+1)}\frac{\partial B_i}{\partial t} = 0$$

$$(4.72)$$

where $\beta^2 = (\gamma^2-1)/\gamma^2$ was used in the last step. The first term in the parentheses in Eq. (4.72) is zero by Eq. (4.70) and the second term in parentheses can be changed to a time derivative using Eq. (4.68). Equation (4.72) then reduces to

$$\varepsilon_{ijk}\frac{\partial E_k}{\partial x_j}+\frac{\gamma}{(\gamma+1)}\frac{1}{c}\frac{\partial B_i}{\partial t}+\frac{1}{(\gamma+1)}\frac{1}{c}\frac{\partial B_i}{\partial t}=0$$

from which

$$\varepsilon_{ijk}\frac{\partial E_k}{\partial x_j}=-\frac{1}{c}\frac{\partial B_i}{\partial t} \tag{4.73}$$

which is Maxwell's equation for the curl of E. That is, Eq. (4.73) is the same as

$$\nabla\times\mathbf{E}=-\frac{1}{c}\frac{\partial \mathbf{B}}{\partial t} \tag{4.74}$$

Equation (4.70) is the same as

$$\nabla\cdot\mathbf{B}=0 \tag{4.75}$$

Equations (4.74) and (4.75) are Maxwell's homogeneous equations.

Both Eqs. (4.75) and (4.74) for the divergence of **B** and the curl of **E** have been derived from Eq. (4.60) for the curl of **E'**. In order to derive other equations by transforming Eq. (4.61), it is necessary to determine how the charge density ρ' transforms under a Lorentz transformation. Since it has been postulated that charge is invariant, then

$$\rho'dV'=\rho dV \tag{4.76}$$

where

$$dV={}_{\perp}dx_1\,{}_{\perp}dx_2\,{}_{\parallel}dx \text{ and } dV'={}_{\perp}dx_1'\,{}_{\perp}dx_2'\,{}_{\parallel}dx' \tag{4.77}$$

Then

$${}_{\perp}dx_1'={}_{\perp}dx_1 \text{ and } {}_{\perp}dx_2'={}_{\perp}dx_2 \tag{4.78}$$

and

$${}_{\parallel}x'=\gamma({}_{\parallel}x-ut)$$
$${}_{\parallel}dx'=\gamma{}_{\parallel}dx \tag{4.79}$$

Substituting Eqs. (4.77), (4.78), and (4.79) into Eq. (4,76), we find that

$$\rho' = \frac{\rho}{\gamma} \tag{4.80}$$

Using Eqs. (4.69) and (4.80), we can write Eq. (4.61) as

$$\frac{\partial E_i'}{\partial x_i} - \frac{(\gamma-1)}{\gamma}\alpha_i\alpha_j\frac{\partial E_i'}{\partial x_j} = \frac{\rho}{\gamma} \tag{4.81}$$

Substituting Eq. (4.56) for E_i' in Eq. (4.81), we can write

$$\alpha_i\alpha_j\frac{\partial E_j}{\partial x_i} - \frac{1}{\gamma\beta}\varepsilon_{ijk}\alpha_j\frac{\partial B_k}{\partial x_i} - \frac{(\gamma-1)}{\gamma}\alpha_i\alpha_j\alpha_i\alpha_k\frac{\partial E_k}{\partial x_j}$$
$$+ \frac{(\gamma-1)}{\gamma^2\beta}\alpha_i\alpha_j\varepsilon_{irs}\alpha_r\frac{\partial B_s}{\partial x_j} = \frac{\rho}{\gamma}$$
$$\alpha_i\alpha_j\frac{\partial E_j}{\partial x_i}\left(1 - \frac{(\gamma-1)}{\gamma}\right) - \frac{1}{\gamma\beta}\alpha_j\varepsilon_{jki}\frac{\partial B_k}{\partial x_i} = \frac{\rho}{\gamma} \tag{4.82}$$

which, by using Eq. (4.68), can be written as

$$-\frac{1}{\beta c}\alpha_j\frac{\partial E_j}{\partial t}\frac{1}{\gamma} - \frac{1}{\gamma\beta}\alpha_j\varepsilon_{jki}\frac{\partial B_k}{\partial x_i} = \frac{\rho}{\gamma}$$
$$-\alpha_i\frac{1}{c}\frac{\partial E_i}{\partial t} + \alpha_i\varepsilon_{ijk}\frac{\partial B_k}{\partial x_j} = \rho\beta \tag{4.83}$$

Multiplying the right-hand side of Eq. (4.83) by $\alpha_i\alpha_i = 1$ and defining the current density of the source charges as $J_i = \rho u\alpha_i = \rho\beta c\alpha_i$, we can write Eq. (4.83) as

$$\alpha_i\left(-\frac{1}{c}J_i - \frac{1}{c}\frac{\partial E_i}{\partial t} + \varepsilon_{ijk}\frac{\partial B_k}{\partial x_j}\right) = 0 \tag{4.84}$$

Since J_i is in the direction of α_i, the total vector in parentheses cannot be perpendicular to α_i. Therefore, the expression in parenthesis in Eq. (4.84) must vanish and

$$\varepsilon_{ijk} \frac{\partial B_k}{\partial x_j} = \frac{1}{c} J_i + \frac{1}{c} \frac{\partial E_i}{\partial t} \tag{4.85}$$

which is Maxwell's equation for the curl of **B** and is the same as

$$\nabla \times \mathbf{B} = \frac{1}{c} \mathbf{J} + \frac{1}{c} \frac{\partial \mathbf{E}}{\partial t} \tag{4.80}$$

We first rewrite Eqs. (4.81) and Eq. (4.57), and then substitute Eq. (4.57) for E_i' in Eq. (4.81). Thus,

$$\frac{\partial E_i'}{\partial x_i} - \frac{(\gamma-1)}{\gamma} \alpha_i \alpha_j \frac{\partial E_i'}{\partial x_j} = \frac{\rho}{\gamma}$$

$$E_i' = \frac{1}{\gamma} E_i + \frac{(\gamma-1)}{\gamma} \alpha_i \alpha_j E_j$$

$$\frac{\partial E_i}{\partial x_i} + (\gamma-1)\alpha_i \alpha_j \frac{\partial E_j}{\partial x_i} - \frac{(\gamma-1)}{\gamma} \alpha_i \alpha_j \frac{\partial E_i}{\partial x_j} - \frac{(\gamma-1)^2}{\gamma} \alpha_i \alpha_j \alpha_i \alpha_k \frac{\partial E_k}{\partial x_j} = \rho$$

$$\frac{\partial E_i}{\partial x_i} + \alpha_i \alpha_j \frac{\partial E_j}{\partial x_i} \left[(\gamma-1) - \frac{(\gamma-1)}{\gamma} - \frac{(\gamma-1)^2}{\gamma} \right] = \rho \tag{4.87}$$

The term in the square bracket in Eq. (4.87) is equal to zero so that Eq. (4.87) reduces to

$$\frac{\partial E_i}{\partial x_i} = \rho \tag{4.88}$$

which is Maxwell's equation for the divergence of **E** and is the same as

$$\nabla \cdot \mathbf{E} = \rho \tag{4.89}$$

Note that Eq. (4.89) is same as Eq. 4.32 that we found for static fields.

We have therefore derived the four Maxwell equations given by Eqs. (4.74), (4.75), (4.86), and (4.89) directly from special relativity and Coulomb's law. These equations, in rationalized Heaviside-Lorentz units, are summarized in Eq. (4.90).

$$\nabla \times \mathbf{E} = -\frac{1}{c}\frac{\partial \mathbf{B}}{\partial t}$$
$$\nabla \cdot \mathbf{B} = 0$$
$$\nabla \times \mathbf{B} = \frac{1}{c}\mathbf{J} + \frac{1}{c}\frac{\partial \mathbf{E}}{\partial t} \tag{4.90}$$
$$\nabla \cdot \mathbf{E} = \rho$$

4.6 Electromagnetic Waves

From Eqs. (4.73) and (4.85), we can write the two Maxwell curl equations for free space (with the current density set to zero) as

$$\varepsilon_{ijk}\frac{\partial E_k}{\partial x_j} = -\frac{1}{c}\frac{\partial B_i}{\partial t} \tag{4.91}$$

$$\varepsilon_{ijk}\frac{\partial B_k}{\partial x_j} = \frac{1}{c}\frac{\partial E_i}{\partial t} \tag{4.92}$$

Rename the subscripts in Eq. (4.92) to

$$\varepsilon_{rsi}\frac{\partial B_i}{\partial x_s} = \frac{1}{c}\frac{\partial E_r}{\partial t} \tag{4.93}$$

Taking the time derivative of both sides of Eq. (4.93) and interchanging the order of differentiation, we obtain

$$\varepsilon_{rsi}\frac{\partial}{\partial x_s}\frac{\partial B_i}{\partial t} = \frac{1}{c}\frac{\partial^2 E_r}{\partial t^2} \tag{4.94}$$

Substituting $\partial B_i/\partial t$ from Eq. (4.91) into Eq. (4.94), we obtain

$$\varepsilon_{irs}\varepsilon_{ijk}\frac{\partial}{\partial x_s}\frac{\partial E_k}{\partial x_j} = -\frac{1}{c^2}\frac{\partial^2 E_r}{\partial t^2}$$

$$\left(\delta_{rj}\delta_{sk} - \delta_{rk}\delta_{sj}\right)\frac{\partial}{\partial x_s}\frac{\partial E_k}{\partial x_j} = -\frac{1}{c^2}\frac{\partial^2 E_r}{\partial t^2}$$

$$\frac{\partial}{\partial x_s}\frac{\partial E_s}{\partial x_r} - \frac{\partial}{\partial x_s}\frac{\partial E_r}{\partial x_s} = -\frac{1}{c^2}\frac{\partial^2 E_r}{\partial t^2} \tag{4.95}$$

Interchanging the order of differentiation in the first term of Eq. (4.95) and recognizing from Eq. (4.88) that in free space ($\rho = 0$) the divergence of E_s is zero ($\partial E_s/\partial x_s = 0$), then the first term in Eq. (4.95) vanishes and Eq. (4.95) reduces to the wave equation

$$\frac{\partial^2 E_r}{\partial x_s \partial x_s} = \frac{1}{c^2}\frac{\partial^2 E_r}{\partial t^2} \tag{4.96}$$

or in symbolic notation

$$\nabla^2 \mathbf{E} = \frac{1}{c^2}\frac{\partial^2 \mathbf{E}}{\partial t^2} \tag{4.97}$$

Eq. (4.96) has solutions of the form

$$E_r = f_r\left(\alpha_i x_i \pm ct\right) \tag{4.98}$$

where the unit vector α_i has the property $\alpha_i\alpha_i = 1$. We can verify that Eq. (4.98) is a solution of Eq. (4.96) by noting that if we write

$$f_r' = \frac{\partial f_r(\xi)}{\partial \xi} \tag{4.99}$$

then $\partial E_r/\partial x_s = f_r'\alpha_s$, $\partial^2 E_r/\partial x_s\partial x_s = f_r''\alpha_s\alpha_s = f_r''$, $\partial E_r/\partial t = \pm cf_r'$, and $\partial^2 E_r/\partial t^2 = c^2 f_r''$ so that Eq. (4.96) is satisfied.

Now consider the solution to the wave equation given by Eq.

(4.98) using the minus sign. At some time t_1, the locus in space of $E_r = f_r(\alpha_i x_i - ct) = \text{constant}$, will be the locus in space of $\alpha_i x_i' - ct_1 = \text{constant}$, which is the equation of a plane with normal α_i. A second plane at t_2 will satisfy the equation $\alpha_i x_i'' - ct_2 = \text{constant}$. If E_r is to have the same value at both values of time, then we must have (see Fig. 4.3)

$$\alpha_i x_i' - ct_1 = \alpha_i x_i'' - ct_2 \text{ or } \alpha_i\left(x_i'' - x_i'\right) = c\left(t_2 - t_1\right) \text{ or } d = c\left(t_2 - t_1\right),$$

where d is the distance between the planes. Thus, $c = d/(t_2 - t_1)$ is the velocity of propagation of the waves. Such waves are called *plane waves*. Thus, Eq. (4.98) is a plane wave solution of Eq. (4.96) in which the velocity of propagation is c. Use of the plus sign in Eq. (4.98) corresponds to the plane waves propagating in the opposite (minus α_i) direction.

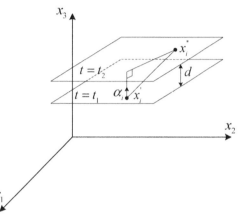

Fig. 4.3 Plane wave

Harmonic Waves

As a specific example of a plane wave solution given by Eq. (4.98), consider the cosine function

$$E_r = A_r \cos k\left(\alpha_i x_i \pm ct\right) = A_r \cos\left(k\alpha_i x_i \pm \omega t\right) \qquad (4.100)$$

where

$$\omega = kc = 2\pi f \qquad (4.101)$$

is the angular frequency in radians per second and f is the frequency in cycles per second, or Hertz.

If $\alpha_i = (1, 0, 0)$, the wave is traveling in the x_1-direction and

$$E_r = A_r \cos\left(kx_1 \pm \omega t\right) \tag{4.102}$$

The value of E_r repeats itself each time kx_1 increases by 2π. Thus, k is the number of waves in a length of 2π and is called the *wave number*. The length of one wave is therefore $2\pi/k$ and is called the *wavelength* λ. That is,

$$\lambda = 2\pi/k \tag{4.103}$$

It is often convenient to represent the harmonic solution in Eq. (1.100) in complex form as

$$E_r = A_r e^{i(k\alpha_i x_i \pm \omega t)} = A_r e^{i2\pi\left(\frac{\alpha_i x_i}{\lambda} \pm ft\right)} \tag{4.104}$$

where the field values are obtained by taking the real part of Eq. (4.104).

Because Eq. (4.104) is a solution of the free space wave equation, the divergence of E_r will be zero. Therefore, we can write

$$E_{r,r} = 0 = A_r ik\alpha_r e^{i(k\alpha_i x_i \pm \omega t)} \tag{4.105}$$

For there to be a wave, k must be nonzero. Therefore, it must be the case that

$$A_r \alpha_r = 0 \tag{4.106}$$

which says that the *E*-field is a transverse wave, perpendicular to the direction of propagation.

Maxwell's Eq. (4.91) is

$$\varepsilon_{ijk} \frac{\partial E_k}{\partial x_j} = -\frac{1}{c}\frac{\partial B_i}{\partial t} \tag{4.107}$$

Substituting Eq. (4.104) in Eq. (4.107), we obtain

$$ik\varepsilon_{ijk}\alpha_j E_k = -\frac{1}{c}\frac{\partial B_i}{\partial t} \tag{4.108}$$

Integrating Eq. (4.108) with respect to time, we obtain

$$\frac{ik}{\pm i\omega}\varepsilon_{ijk}\alpha_j F_{ik} = -\frac{1}{c}B_i$$

or

$$B_i = \mp\varepsilon_{ijk}\alpha_j E_k \qquad (4.109)$$

where Eq. (4.101) was used.

Eq. (4.109) says that B_i is also a transverse wave and is also perpendicular to E_k. The upper minus sign in Eq. (4.109) corresponds to the plus sign in Eq. (4.104); i.e., wave propagation in the minus α_i direction. The waves always propagate in the direction of $\mathbf{E}\times\mathbf{B}$.

4.7 Quaternion Representation of Maxwell's Equations

It is possible to combine the four Maxwell Equations in Eq. (4.90) into a single quaternion equation. To do this, we can first write the two curl equations as

$$\nabla\times\mathbf{E}+\frac{1}{c}\frac{\partial\mathbf{B}}{\partial t}=0 \qquad (4.110)$$

and

$$\nabla\times\mathbf{B}-\frac{1}{c}\frac{\partial\mathbf{E}}{\partial t}=\frac{1}{c}\mathbf{J} \qquad (4.111)$$

If we multiply Eq. (4.110) by the imaginary unit vector i, and add the resulting equation to Eq. (4.111), we obtain

$$\nabla\times(\mathbf{B}+i\mathbf{E})-\frac{1}{ic}\frac{\partial(\mathbf{B}+i\mathbf{E})}{\partial t}=\frac{1}{c}\mathbf{J} \qquad (4.112)$$

We can now write the two Maxwell divergence equations from Eq. (4.90) as

$$\nabla\cdot\mathbf{B}=0 \qquad (4.113)$$

and

$$\nabla \cdot \mathbf{E} = \rho \tag{4.114}$$

Multiplying Eq. (4.114) by i and adding the resulting equation to Eq. (4.113), we obtain

$$\nabla \cdot (\mathbf{B} + i\mathbf{E}) = i\rho \tag{4.115}$$

Let

$$\psi = \mathbf{B} + i\mathbf{E} \tag{4.116}$$

Subtracting Eq. (4.115) from Eq. (4.112), and using Eq. (4.116), we can write

$$\frac{i}{c}\frac{\partial \psi}{\partial t} - \nabla \cdot \psi + \nabla \times \psi = -i\rho + \frac{1}{c}\mathbf{J} \tag{4.117}$$

We can consider ψ in Eq. (4.116) to be a complex quaternion in which $\mathbf{B}$ and $\mathbf{E}$ are pure quaternions with zero scalar values (see Appendix A). If we then define D to be the quaternion differential operator

$$D = \left(\frac{i}{c}\frac{\partial}{\partial t} + \nabla \right) \tag{4.118}$$

and

$$C = -i\rho + \frac{1}{c}\mathbf{J} \tag{4.119}$$

to be the current quaternion, then using the rules of quaternion multiplication (see Appendix A), Eq. (4.117) reduces to

$$D\psi = C \tag{4.120}$$

Thus, the four Maxwell equations in Eq. (4.90) reduce to the simple, single equation given by Eq. (4.120). Note that the electric and magnetic fields only appear together as the complex quaternion ψ given in Eq. (4.116).

Chapter 5

An Alternate Theory of Gravitation

In my novel, *Peggy's Discovery*, Peggy, as a college freshman, watches her engineering professor uncle go through the derivation of Maxwell's equations from Coulomb's law and special relativity given in Chapter 4. She then realizes that the same derivation will work if you just replace Coulomb's law with Newton's law of gravitation, which she had covered in her freshman physics class. She wonders why Einstein didn't just do this when trying to get gravitation to agree with special relativity, instead of spending a decade developing general relativity complete with 4-dimensional curved space.

The answer is that Einstein did not know about this derivation. In a sense, he invented special relativity to agree with Maxwell's equations by postulating that the speed of light is a constant, independent of the velocity of the source. In this chapter, we will show what Einstein would have found if he carried out the derivation in Chapter 4 using Newton's law of gravitation. We will see that it leads to four equations, exactly analogous to Maxwell's equations, with two gravitational fields analogous to the electric and magnetic fields.

It turns out that Oliver Heaviside published a paper in *The Electrician* in 1893 entitled, "A Gravitational and Electromagnetic Analogy," in which he suggests that the four equations we derive in this chapter [see Eq. (5.37)] could very well be a more complete theory of gravitation. This was twelve years before Einstein published his special theory of relativity. There is some indication that Einstein may have been familiar with Heaviside's paper, but being only an analogy with no apparent theoretical underpinning, Einstein took a completely different route in pursuit of a theory of gravitation, and ended up with the curved space-time of his general theory of relativity. Interestingly, a certain linear approximation of Einstein's field equation for weak fields and flat space, leads to the "analogous" gravitational Maxwell's equations of Eq. (5.37). These equations describe what has become known as gravito-electromagnetism (GEM). The derivation in this chapter shows that these GEM equations follow directly from Newton's law of gravitation and special relativity.

5.1 Newton's Law of Gravitation

Recall that Coulomb's law is given in Eq. (4.1) as

$$\mathbf{F}_e = k_e \frac{q_1 q_2}{r_{12}^2} \hat{\mathbf{r}}_{12} \tag{5.1}$$

where $\mathbf{F}_e$ is the force on a charge q_2, located at $\mathbf{r}_2$, due to a charge q_1, located at $\mathbf{r}_1$, and $\hat{\mathbf{r}}_{12}$ is a unit vector in the direction of $\mathbf{r}_2 - \mathbf{r}_1$ and $r_{12} = |\mathbf{r}_2 - \mathbf{r}_1|$. The constant k_e depends on the units used.

Newton's law of gravitation can be written as

$$\mathbf{F}_g = -G \frac{m_1 m_2}{r_{12}^2} \hat{\mathbf{r}}_{12} \tag{5.2}$$

which gives the force on mass m_2 due to mass m_1. The similarity of Eqs. (5.1) and (5.2) is obvious. The charges in Eq. (5.1) can be either positive or negative, while the masses in Eq. (5.2) are always positive. The minus sign in Eq. (5.2) indicates that the gravitational force is always an attractive force. Recalling that force has units of MLT^{-2}, the gravitational constant, G, in Eq. (5.2) will have units of $\text{M}^{-1}\text{L}^3\text{T}^{-2}$. In SI units, $G = 6.67 \times 10^{-11} \text{ N-}(m/kg)^2$, and in CGS units, $G = 6.67 \times 10^{-8} \text{ g}^{-1}\text{cm}^3\text{s}^{-2}$.

Following our definition of the electric field in Eq. (4.23), we will define the gravitational field $\mathbf{R}$ due to a mass M to be the force per unit mass on a test mass m. Then, from Eq. (5.2), we can write

$$\mathbf{R} = \frac{\mathbf{F}_g}{m} = -\frac{GM}{r^2} \boldsymbol{\alpha} \tag{5.3}$$

where r is the distance between M and m, and $\boldsymbol{\alpha}$ is a unit vector in the direction of m from M.

Following our discussion of the electric field in Section 4.3, we define the gravitational flux Ψ_G on a closed surface S as

$$\Psi_G = \oiint_S \mathbf{R} \cdot \mathbf{n} \, dA \tag{5.4}$$

where $\mathbf{n}$ is an outward unit vector normal to the area dA. The gravitational field, $\mathbf{R}$, will look like the electric field in Fig. 4.2, except all the arrows will be pointing inward. The unit vectors $\boldsymbol{\alpha}$ and $\mathbf{n}$ in Eqs. (5.3) and (5.4) will be parallel, and therefore, from Eq. (5.3), we can write

$$\mathbf{R} \cdot \mathbf{n} = -\frac{GM}{r^2} \tag{5.5}$$

Therefore, the gravitational flux Ψ_G in Eq. (5.4), can be written as

$$\Psi_G = \oiint_S \mathbf{R} \cdot \mathbf{n} \, dA = -\oiint_S \frac{GM}{r^2} dA = -4\pi GM \tag{5.6}$$

since the area of the sphere in Fig. 4.2 is $4\pi r^2$. Therefore, the total gravitational flux leaving the closed spherical surface is negative, and proportional to the total mass within the closed surface.

More generally, we can consider a mass density ρ_m such that the mass M within a small volume element ΔV is given by

$$M = \rho_m \Delta V \tag{5.7}$$

The *divergence* of the vector $\mathbf{R}$ is defined by (see Eq. (4.28))

$$\nabla \bullet \mathbf{R} = \lim_{\Delta V \to 0} \frac{\Psi_G}{\Delta V} = \lim_{\Delta V \to 0} \frac{1}{\Delta V} \oiint_S \mathbf{R} \cdot \mathbf{n} \, dA \tag{5.8}$$

from which, using Eqs. (5.6) and (5.7), we can write

$$\nabla \bullet \mathbf{R} \Delta V = -4\pi G \rho_m \Delta V \tag{5.9}$$

Taking the limit as $\Delta V \to 0$ and integrating over the volume enclosed by the closed surface S, we can write

$$\oiiint_V \nabla \cdot \mathbf{R} \, dV = -\oiiint_V 4\pi G \rho_m \, dV \qquad (5.10)$$

or

$$\oiiint_V (\nabla \cdot \mathbf{R} + 4\pi G \rho_m) \, dV = 0 \qquad (5.11)$$

Since the volume V is arbitrary, the integrand in Eq. (5.11) must be zero. We therefore conclude that

$$\nabla \cdot \mathbf{R} = -4\pi G \rho_m \qquad (5.12)$$

which is analogous to the first of Maxwell's equations, given in Eq. (4.23) as

$$\nabla \cdot \mathbf{E} = \rho \qquad (5.13)$$

Gravitational Potential

Let mass M be at the origin and define the gravitational scalar potential

$$\varphi_m = -\frac{GM}{r} \qquad (5.14)$$

If we take the gradient of φ_m in spherical coordinates we can write

$$\nabla \varphi_m = -GM\boldsymbol{\alpha} \frac{d}{dr}\left(\frac{1}{r}\right) = \frac{GM}{r^2}\boldsymbol{\alpha} \qquad (5.15)$$

Comparing Eqs. (5.3) and (5.15) we see that we can write $\mathbf{R}$ as

$$\mathbf{R} = -\nabla \varphi_m \qquad (5.16)$$

Because $\mathbf{R}$ can be written as the negative gradient of a scalar, it follows that

$$\nabla \times \mathbf{R} = 0 \qquad (5.17)$$

since the curl of a gradient is always zero.

We therefore see from Newton's law of gravitation that when the masses are at rest the gravitational field **R** satisfies the gravitational static equations (5.17) and (5.12), which can be written in component form as

$$\varepsilon_{ijk} \frac{\partial R_k}{\partial x_j} = 0 \tag{5.18}$$

and

$$\frac{\partial R_i}{\partial x_i} = -4\pi G \rho_m \tag{5.19}$$

Note that the subscript m in Eq. (5.19) is not a vector index; it just denotes mass density.

5.2 A Gravitational Lorentz Force

Let a collection of source masses with a mass density $\rho_m{}'$ be at rest in a reference frame S' which has velocity $u_i = u\alpha_i$ and moves uniformly relative to another reference frame S. This mass density gives rise to a static gravitational field $R_i{}'$ in the frame S' and will produce a force $F_i{}' = m'R_i{}'$ on a test mass m'. The force on a test mass m as measured by observers in reference frame S can be found by using the force transformation given by Eq. (3.40).

Two cases can be distinguished. If the velocity v_i of the test mass is zero, the gravitational field R_i measured in S is defined by the force relation $F_i = mR_i$. The masses m and m' are rest masses. Assuming rest mass is invariant (i.e., $m = m'$) and setting $F_i = mR_i$, $F_j{}' = m'R_j{}'$ and $v_j = 0$ in Eq. (3.40), we find that R_i is given by

$$R_i = \left[\gamma\delta_{ij} - (\gamma-1)\alpha_i\alpha_j \right] R_j{}' \tag{5.20}$$

If the test mass m is now allowed to move with a velocity v_i then Eq. (3.40) becomes

$$F_i = \left[\gamma \delta_{ij} - (\gamma - 1)\alpha_i\alpha_j \right] mR_j{}' + \frac{m\gamma\beta}{c}\left(\alpha_i R_j{}' - \alpha_j R_i{}' \right)v_j$$

which, using Eq. (5.20), can be written as

$$F_i = mR_i + \frac{m}{c}C_{ij}v_j \qquad (5.21)$$

where

$$C_{ij} = \gamma\beta\left(\alpha_i R_j{}' - \alpha_j R_i{}' \right) \qquad (5.22)$$

Since C_{ij} is anti-symmetric, an axial vector S_i, which we will call a gravitational magnetic field, can be defined by the relation

$$C_{ij} = \varepsilon_{ijk}S_k \qquad (5.23)$$

Therefore, Eq. (5.21) can be written as

$$F_i = m\left(R_i + \frac{1}{c}\varepsilon_{ijk}v_j S_k \right) \qquad (5.24)$$

or

$$\mathbf{F} = m\left(\mathbf{R} + \frac{1}{c}\mathbf{v}\times\mathbf{S} \right) \qquad (5.25)$$

which we will call the *Gravitational Lorentz force*. Note that it is directly analogous to the Lorentz force involving electric and magnetic fields given by Eq. (4.50).

This means that, in addition to the normal gravitational field, $\mathbf{R}$, given by Eq. (5.3), there is a second gravitational field, $\mathbf{S}$, analogous to the magnetic field, which is generated by moving masses, and exerts a force only on other moving masses. We will return to the significance of this second gravitational field later in this chapter.

5.3 Gravitational Version of Maxwell's Equations

The derivation of a gravitational version of Maxwell's equations follows directly from the derivation given in Section 4.5. In this section, we will indicate where the differences occur. From Eqs.

(5.18) and (5.19), we can write the static gravitational equations for the static field, R_i' in the S' reference frame as

$$\varepsilon_{ijk} \frac{\partial R_k'}{\partial x_j'} = 0 \tag{5.26}$$

and

$$\frac{\partial R_i'}{\partial x_i'} = -4\pi G \rho_m' \tag{5.27}$$

which are analogous to Eqs. (4.60) and (4.61).

The first part of the derivation in Section 4.5 proceeds with no changes, simply replacing the fields, E_i and B_i with R_i and S_i, respectively. Thus, Eq. (4.70) becomes

$$\frac{\partial S_i}{\partial x_i} = 0 \tag{5.28}$$

which is the same as

$$\nabla \bullet \mathbf{S} = 0 \tag{5.29}$$

and Eq. (4.73) becomes

$$\varepsilon_{ijk} \frac{\partial R_k}{\partial x_j} = -\frac{1}{c} \frac{\partial S_i}{\partial t} \tag{5.30}$$

which is the gravitational Maxwell's equation for the curl of $\mathbf{R}$. That is, Eq. (5.29) is the same as

$$\nabla \times \mathbf{R} = -\frac{1}{c} \frac{\partial \mathbf{S}}{\partial t} \tag{5.31}$$

From Eq. (5.27), we see that as we continue the derivation, we must include the factor $-4\pi G$ in terms involving the mass density. Thus, Eq. (4.81) will become

$$\frac{\partial R_i'}{\partial x_i} - \frac{(\gamma-1)}{\gamma}\alpha_i\alpha_j\frac{\partial R_i'}{\partial x_j} = -\frac{4\pi G\rho_m}{\gamma} \tag{5.32}$$

and Eq. (4.83) will become

$$-\alpha_i\frac{1}{c}\frac{\partial R_i}{\partial t} + \alpha_i\varepsilon_{ijk}\frac{\partial S_k}{\partial x_j} = -4\pi G\rho_m\beta \tag{5.33}$$

Multiplying the right-hand side of Eq. (5.33) by $\alpha_i\alpha_i = 1$ and defining the gravitational current density of the source masses as $K_i = \rho_m u\alpha_i = \rho_m\beta c\alpha_i$, we can write Eq. (5.33) as

$$\alpha_i\left(\frac{4\pi G}{c}K_i - \frac{1}{c}\frac{\partial R_i}{\partial t} + \varepsilon_{ijk}\frac{\partial S_k}{\partial x_j}\right) = 0 \tag{5.34}$$

Since K_i is in the direction of α_i, the total vector in parentheses cannot be perpendicular to α_i. Therefore, the expression in parenthesis in Eq. (5.34) must vanish and

$$\varepsilon_{ijk}\frac{\partial S_k}{\partial x_j} = -\frac{4\pi G}{c}K_i + \frac{1}{c}\frac{\partial R_i}{\partial t} \tag{5.35}$$

which is the gravitational Maxwell's equation for the curl of **S** and is the same as

$$\nabla\times\mathbf{S} = -\frac{4\pi G}{c}\mathbf{K} + \frac{1}{c}\frac{\partial\mathbf{R}}{\partial t} \tag{5.36}$$

We have therefore derived the four gravitational Maxwell equations given by Eqs. (5.12), (5.29), (5.31), and (5.36) directly from special relativity and Newton's law of gravitation.

These equations, in CGS units, are summarized in Eq. (5.37).

$$\nabla \times \mathbf{R} = -\frac{1}{c}\frac{\partial \mathbf{S}}{\partial t}$$

$$\nabla \bullet \mathbf{S} = 0$$

$$\nabla \times \mathbf{S} = -\frac{4\pi G}{c}\mathbf{K} + \frac{1}{c}\frac{\partial \mathbf{R}}{\partial t}$$

$$\nabla \bullet \mathbf{R} = -4\pi G \rho_m$$

(5.37)

What does it mean that gravity must satisfy the four Maxwell-like equations in Eqs. (5.37)? First of all, in free space, in the absence of masses, $((\rho_m = 0; \mathbf{K} = 0)$, Eqs. (5.37) yield a wave equation with solutions being transverse waves traveling at the speed of light. These gravitational waves will be just like the electromagnetic waves described in Section 4.6, this time involving the two fields $\mathbf{R}$ and $\mathbf{S}$. They would be produced by accelerating masses, for example, binary stars.

The equations predict that moving masses produce $\mathbf{S}$ fields, and these $\mathbf{S}$ fields exert forces on other moving masses. The gravitational Lorentz force is given by Eq. (5.25) as

$$\mathbf{F} = m\left(\mathbf{R} + \frac{1}{c}\mathbf{v} \times \mathbf{S}\right)$$

(5.38)

From Eq. (4.55), we see that the $\mathbf{S}$ field can be expressed in terms of the $\mathbf{R}$ field by the equation

$$\mathbf{S} = \frac{1}{c}\mathbf{u} \times \mathbf{R}$$

(5.39)

Substituting Eq. (5.39) into Eq. (5.38), we can write the gravitational Lorentz force as

$$\mathbf{F} = m\left(\mathbf{R} + \frac{1}{c^2}\mathbf{v} \times (\mathbf{u} \times \mathbf{R})\right)$$

(5.40)

We also know from Eq. (4.53) that

$$\mathbf{\alpha \cdot R} = \mathbf{\alpha \cdot R'} \tag{5.41}$$

so that the component of $\mathbf{R}$ in the direction of motion of the mass is invariant. Also, from Eq. (4.54) we know that

$$\mathbf{\alpha \times R} = \gamma \mathbf{\alpha \times R'} \tag{5.42}$$

so that the component of $\mathbf{R}$ perpendicular to the direction of motion of the mass is larger than the corresponding component of $\mathbf{R'}$ by the factor γ.

From Eq. (5.39), we see that the value of the $\mathbf{S}$ field generated by a moving mass is zero in the direction of motion of the mass, and is maximum perpendicular to the direction of motion.

We can rewrite Eq. (5.40) as

$$\mathbf{F} = \mathbf{F}_R + \mathbf{F}_S \tag{5.43}$$

where

$$\mathbf{F}_R = m\mathbf{R} \tag{5.44}$$

and

$$\mathbf{F}_S = \frac{m}{c^2} \mathbf{v} \times (\mathbf{u} \times \mathbf{R}) \tag{5.45}$$

The force $\mathbf{F}_R$ is the normal gravitational force, and the new force $\mathbf{F}_S$ is a force on moving masses due to other moving masses, analogous to magnetic field forces.

Suppose that a mass, M, is moving with a velocity, $\mathbf{u}$. The direction of the gravitational field vector, $\mathbf{R}$, perpendicular to $\mathbf{u}$, will be pointing toward the mass, M. Thus, from Eq. (5.45), the force $\mathbf{F}_S$ on the mass, m, will be a repulsive force if the velocity, $\mathbf{v}$, of the mass, m, is in the direction of $\mathbf{u}$, and it will be an attractive force if the two masses, M and m, are moving in opposite directions. Note that this is the opposite from the way parallel, current carrying conductors attract and repel.

As can be seen from Eqs. (5.44) and (5.45), the force $\mathbf{F}_S$ is much smaller than the force $\mathbf{F}_R$ for mass velocities much less than the speed of light.

Chapter 6

Toward a Unified Field Theory

In the preface, I tell the story of how I came to obtain a copy of the book *A Unified Field Theory* by Miles V. Hayes over forty years ago, and how, for forty years, I never read it. This puts me in good company, for this book seems to have all but disappeared. The author gave copies to about 30 libraries throughout the United States and about 50 copies to libraries in 35 countries around the world. It seems that few of these copies have ever been read.

The book contains 28 short chapters tucked into only 70 pages, but the breadth of the topics covered is breathtaking; how a single equation can explain all of physical reality. In this chapter, I will explain the more technical parts of that book, but you will need to find a copy to read about the author's more philosophical musings.

6.1 The Field Equation

The Preface of the Hayes book is called a Summary and the first paragraph reads:

> The universe consists of a complex quaternionic field which is a function of space-time such that, in the sense defined by the algebra of complex quaternions, its rate of change is proportional to the square of its magnitude.

In the next paragraph is a single equation, which the author calls the field equation and which he asserts describes all of physical reality.

On page 1, Hayes, who was an associate professor of engineering at Dartmouth when he wrote the book in 1964, explains how he came to write the book. He says he came up with the theory at the end of 1961 and then writes

> In the following year I developed the theory and submitted three papers to the Physical Review. They were

rejected on the grounds that the theory is 'speculative'. Had all scientists refrained from speculation there would be no physics and no Physical Review. I have decided to publish the theory privately.

The difficulty in gaining any acceptance or even consideration for the theory is not, I think, that it fails to explain physics but that it explains too much too simply. It seems incredible that a short simple equation $D\psi = \frac{1}{2}$ $\psi^* \iota \psi$ should be adequate to explain all of physical reality. I don't blame physicists for being skeptical, since I have a great deal of trouble believing it myself

But I have even more trouble believing the theory is false when it explains so much. What other theory is proposed that better explains, or indeed explains at all, the existence of classical electromagnetism, classical mechanics, quantum mechanics, electric charge, matter and anti-matter, special relativity, the conservation laws, and promises to explain elementary particles? I am open to conversion, but until someone shows that the theory leads to fallacious results, which no one has done, or someone produces a simpler and broader theory, which no one has done, I propose to accept on a tentative basis the unified field theory, and I recommend that others at least consider doing likewise.

On page 52, Hayes tells how he came to develop the field equation when he writes

I was first led to the field equation by the following line of thought. My point of departure was a scientific hypothesis or artistic feeling or religious faith that the universe is both unified and simple. Physics has reduced it to fields and elementary particles, but this is dualistic, not unified. Fields can exist, as empty space, without particles, but particles cannot exist without space, which is a field, in which to exist, so fields are more fundamental than particles. A unified theory of physics therefore should attempt to explain particles in terms of a field.

Hayes was familiar with L. Silberstein's book *The Theory of Relativity*, published in 1914, in which the author develops the quaternion representation of Maxwell's equations as we presented them in Section 4.7. Recall from Eqs. (4.116), (4.118) – (4.120) that we can write Maxwell's equations as the single equation

$$D\psi = C \tag{6.1}$$

where

$$D = \left(\frac{i}{c} \frac{\partial}{\partial t} + \nabla \right) \tag{6.2}$$

$$\psi = \mathbf{B} + i\mathbf{E} \tag{6.3}$$

and

$$C = -i\rho + \frac{1}{c}\mathbf{J} \tag{6.4}$$

Hayes uses **H** instead of **B** for the magnetic field vector and he uses natural units in which $c = 1$.

Hayes postulated that *particles are standing light waves,* but realized that Maxwell's equations are linear and therefore could not account for stable standing waves in the absence of material boundaries. He therefore realized that his field equation must be nonlinear. He needed to make the right-hand side of Eq. (6.1) nonlinear, so he postulated that the charge density was some function of the electromagnetic field energy density and he recognized that the current density could be the Poynting flux **E**×**B**. After playing around for awhile, he came up with the following nonlinear field equation:

$$D\psi = \frac{1}{2}\psi *\iota\psi \tag{6.5}$$

where

$$D = \iota\frac{\partial}{\partial t} + \nabla = \iota\frac{\partial}{\partial t} + \mathbf{i}\frac{\partial}{\partial x} + \mathbf{j}\frac{\partial}{\partial y} + \mathbf{k}\frac{\partial}{\partial z} \tag{6.6}$$

and

$$\psi = \mathbf{B} + \iota\mathbf{E} \tag{6.7}$$
$$\psi^* = \mathbf{B} - \iota\mathbf{E} \tag{6.8}$$
$$\iota\iota = \mathbf{ii} = \mathbf{jj} = \mathbf{kk} = -1 \tag{6.9}$$
$$\iota i = i\iota, \; \iota j = j\iota, \; \iota k = k\iota \tag{6.10}$$
$$\mathbf{ij} = -\mathbf{ji} = \mathbf{k}, \; \mathbf{jk} = -\mathbf{kj} = \mathbf{i}, \; \mathbf{ki} = -\mathbf{ik} = \mathbf{j} \tag{6.11}$$

In these equations, ι (iota) is the ordinary, imaginary unit of two-dimensional complex numbers, and is commutative, while $\mathbf{i}$, $\mathbf{j}$, and $\mathbf{k}$ are the ordinary, non-commutative unit vectors of three-dimensional space used in quaternions (see Appendix A), with values equal to the square root of minus one. Thus, ψ is a complex quaternion in which both $\mathbf{B}$ and $\mathbf{E}$ are pure quaternions with zero scalar parts (see Section A.6 of Appendix A).

If we substitute Eqs. (6.6), (6.7), and (6.8) into Eq. (6.5) and carry out the quaternion multiplications (see Eq. (A.15)), we obtain

$$\left(\iota \frac{\partial}{\partial t} + \nabla \right)(\mathbf{B} + \iota \mathbf{E}) = \frac{1}{2}(\mathbf{B} - \iota \mathbf{E})\iota(\mathbf{B} + \iota \mathbf{E}) \qquad (6.12)$$

or

$$\iota \frac{\partial \mathbf{B}}{\partial t} - \frac{\partial \mathbf{E}}{\partial t} - \nabla \bullet \mathbf{B} + \nabla \times \mathbf{B} - \iota \nabla \bullet \mathbf{E} + \iota \nabla \times \mathbf{E}$$

$$= \frac{1}{2}\iota(-\mathbf{B} \bullet \mathbf{B} + \mathbf{B} \times \mathbf{B} - \mathbf{E} \bullet \mathbf{E} + \mathbf{E} \times \mathbf{E}$$

$$-\iota \mathbf{B} \bullet \mathbf{E} + \iota \mathbf{B} \times \mathbf{E} + \iota \mathbf{E} \bullet \mathbf{B} - \iota \mathbf{E} \times \mathbf{B}) \qquad (6.13)$$

from which

$$-\nabla \bullet \mathbf{B} - \iota \nabla \bullet \mathbf{E} + \left(\nabla \times \mathbf{B} - \frac{\partial \mathbf{E}}{\partial t} \right) + \iota \left(\nabla \times \mathbf{E} + \frac{\partial \mathbf{B}}{\partial t} \right)$$

$$= -\frac{1}{2}\iota\left(B^2 + E^2 \right) + \mathbf{E} \times \mathbf{B} \qquad (6.14)$$

Separating the real scalar, imaginary scalar, real vector, and imaginary vector terms, Eq. (6.14) expands to the following four equations:

$$\nabla \bullet \mathbf{B} = 0 \qquad (6.15)$$

$$\nabla \bullet \mathbf{E} = \frac{1}{2}\left(B^2 + E^2 \right) \qquad (6.16)$$

$$\nabla \times \mathbf{B} - \frac{\partial \mathbf{E}}{\partial t} = \mathbf{E} \times \mathbf{B} \qquad (6.17)$$

$$\nabla \times \mathbf{E} + \frac{\partial \mathbf{B}}{\partial t} = 0 \qquad (6.18)$$

Comparing Eqs. (6-15) – (6.18) with Eqs. (4.90), we see that these are Maxwell's four equations where the charge density ρ has been replaced with the electromagnetic energy density $\frac{1}{2}\left(B^2 + E^2\right)$ and the current density **J** has been replaced with the electromagnetic momentum $\mathbf{E} \times \mathbf{B}$.

The right-hand side of Eq. (16.6) is always positive, while the charge density in Maxwell's equations can be positive or negative. This bothered Hayes until he realized that if he changed the sign of ψ in Eq. (6.5), the left-hand side $\mathbf{D}\psi$ changes sign, while the right-hand side does not. Thus, the right-hand side only appears to always be positive, and there is another set of solutions using a negative sign. Hayes also realized that the meaning of changing the sign of ψ in Eq. (6.7) is to change the sign of both **E** and **B**. Doing this will not change the sign of $\mathbf{E} \times \mathbf{B}$ in Eq. (6.17). However, the sign of $\mathbf{E} \times \mathbf{B}$ in Eq. (6.17) must become negative if the **B** on the left-hand side changes sign. This means that $\mathbf{E} \times \mathbf{B}$ must obey a left-hand rule, rather than a right-hand rule. This is equivalent to a flow of negative charges, corresponding to a negative current. Thus, the field equation uncovers the source of positive and negative charges in the handedness of the field vectors.

Eqs. (6.15) and (6.16) are scalar equations and Eqs. (6.17) and (6.18) are vector equations. These four equations therefore expand into the following eight equations in terms of the components of **B** and **E**.

$$\frac{\partial B_x}{\partial x} + \frac{\partial B_y}{\partial y} + \frac{\partial B_z}{\partial z} = 0 \tag{6.19}$$

$$\frac{\partial E_x}{\partial x} + \frac{\partial E_y}{\partial y} + \frac{\partial E_z}{\partial z} = \frac{1}{2}\left(E_x^2 + E_y^2 + E_z^2 + B_x^2 + B_y^2 + B_z^2\right) \tag{6.20}$$

$$\frac{\partial B_z}{\partial y} - \frac{\partial B_y}{\partial z} - \frac{\partial E_x}{\partial t} = E_y B_z - E_z B_y \tag{6.21}$$

$$\frac{\partial B_x}{\partial z} - \frac{\partial B_z}{\partial x} - \frac{\partial E_y}{\partial t} = E_z B_x - E_x B_z \tag{6.22}$$

$$\frac{\partial B_y}{\partial x} - \frac{\partial B_x}{\partial y} - \frac{\partial E_z}{\partial t} = E_x B_y - E_y B_x \qquad (6.23)$$

$$\frac{\partial E_z}{\partial y} - \frac{\partial E_y}{\partial z} + \frac{\partial B_x}{\partial t} = 0 \qquad (6.24)$$

$$\frac{\partial E_x}{\partial z} - \frac{\partial E_z}{\partial x} + \frac{\partial B_y}{\partial t} = 0 \qquad (6.25)$$

$$\frac{\partial E_y}{\partial x} - \frac{\partial E_x}{\partial y} + \frac{\partial B_z}{\partial t} = 0 \qquad (6.26)$$

In his book, Hayes proposed to solve these eight coupled, nonlinear, partial differential equations numerically by replacing them with finite difference equations and iterating them with selected initial conditions and zero boundary conditions. The main problem he faced was that computers fifty years ago didn't have the capacity to solve a four-dimensional mesh (x, y, z, and t) with, say, 100 points on a side. That is no longer a problem today. However, his other problem of selecting initial conditions that might lead to standing wave solutions, which represent physically meaningful solutions such as elementary particles, remains. His idea was that elementary particles would be discrete, stable, oscillatory limit cycles in the nonlinear field.

If the right-hand side of the field equation given by Eq. (6.5) is replaced by a constant, then it reduces to Maxwell's equations given by Eq. (6.1). Thus, Maxwell's equations are an approximation to the field equation in Eq. (6.5). The right-hand side of Eq. (6.5) is quadratic in ψ. Suppose the right-hand side of Eq. (6.5) is set to a linear function of ψ, as in

$$D\psi = K\psi \qquad (6.27)$$

Hayes claims that this equation is equivalent to the Dirac equation of relativistic quantum mechanics, properly interpreted. We will explore this in the next section.

6.2 The Dirac Equation

Fig. 6.1 shows the inscription on a memorial stone in the nave of Westminster Abbey in London, which includes the Dirac equation.

Fig. 6.1 Inscription on stone in Westminster Abbey

Note the similarity of Eq. (6.27) with the equation on the inscription in Fig. 6.1. What is the meaning of the Dirac equation in Fig. 6.1 and where did it come from?

Hayes claims that his field equation given by Eq. (6.5) is consistent with the fundamental postulate of quantum mechanics. From Eq. (6.6) and the expansion of the right-hand side of the field equation given by Eq. (6.14), we can write

$$\left(\imath \frac{\partial}{\partial t} + \nabla \right) \psi = -\frac{1}{2} \imath \left(B^2 + E^2 \right) + \mathbf{E} \times \mathbf{B} \tag{6.28}$$

which, in words says

$$\left(\imath \frac{\partial}{\partial t} + \nabla \right) \psi = \text{scalar energy} + \text{vector momentum} \tag{6.29}$$

This is the fundamental postulate of quantum mechanics, which says that the scalar energy and vector momentum of a differential volume element can be replaced by the differential operators $\imath \partial / \partial t$ and ∇ operating on the wave function ψ.

In particular, wave equations in quantum mechanics are obtained by taking the classical expression for total energy and making the following replacements:

1. Replace momentum $\mathbf{p}$ with $\dfrac{\hbar}{i}\nabla$ (6.30)

2. Replace total energy W with $-\dfrac{\hbar}{i}\dfrac{\partial}{\partial t}$ (6.31)

where $\hbar = h/2\pi$ is the reduced Planck's constant, and the wave function ψ is used as the operand.

For example, to obtain the one-dimensional Schrödinger equation, write the total energy W as the sum of kinetic energy and potential energy.

$$W = \frac{1}{2}mv_x^2 + V(x) \tag{6.32}$$

Using the momentum, $p_x = mv_x$, we can rewrite Eq. (6.32) as

$$W = \frac{p_x^2}{2m} + V(x) \tag{6.33}$$

Applying the replacements given by Eqs. (6-30)-(6.31), we obtain the one-dimensional Schrödinger equation

$$-\frac{\hbar}{i}\frac{\partial \psi(x,t)}{\partial t} = -\frac{\hbar^2}{2m}\frac{\partial \psi(x,t)}{\partial x^2} + V(x)\psi(x,t) \tag{6.34}$$

The Schrödinger equation is a non-relativistic equation that does not explain the fine structure of the hydrogen atom spectrum due to electron spin. Dirac obtained a relativistic wave equation that explained the hydrogen atom fine structure.

The relativistic equation for total energy was found in Eq. (3.20) in Chapter 3 to be given by

$$E^2 = p^2c^2 + m^2c^4 \tag{6.35}$$

Changing the symbol for energy from E to W, we can rewrite Eq. (6.35) as

$$W^2 = c^2\left(p^2 + m^2c^2\right) \tag{6.36}$$

or, in terms of rectangular coordinates,

$$W^2 = c^2 \left(p_x^2 + p_y^2 + p_z^2 + m^2 c^2 \right) \tag{6.37}$$

or

$$W = \pm c \sqrt{p_x^2 + p_y^2 + p_z^2 + m^2 c^2} \tag{6.38}$$

Suppose that we try making the replacements given by Eqs. (6-30)-(6.31), by making the following substitutions in Eq. (6.38).

$$p_x \to \frac{\hbar}{i} \frac{\partial \psi}{\partial x} \qquad p_y \to \frac{\hbar}{i} \frac{\partial \psi}{\partial y}$$

$$p_z \to \frac{\hbar}{i} \frac{\partial \psi}{\partial z} \qquad W \to -\frac{\hbar}{i} \frac{\partial \psi}{\partial t} \tag{6.39}$$

This leads to the following equation:

$$-\frac{\hbar}{i} \frac{\partial \psi}{\partial t} = \pm c \left[\sqrt{-\hbar^2 \left(\frac{\partial^2}{\partial x^2} + \frac{\partial^2}{\partial y^2} + \frac{\partial^2}{\partial z^2} \right) + m^2 c^2} \right] \psi$$

What can this square root mean? How can we get rid of it? In Eq. (6.37) can $\left(p_x^2 + p_y^2 + p_z^2 + m^2 c^2 \right)$ be a perfect square to get rid of the square root?

This is what Dirac tried to do by just writing it out as follows.

$$\left(p_x^2 + p_y^2 + p_z^2 + m^2 c^2 \right) = \left(\alpha_x p_x + \alpha_y p_y + \alpha_z p_z + \beta mc \right)^2$$

$$= \alpha_x^2 p_x^2 + \alpha_x \alpha_y p_x p_y + \alpha_x \alpha_z p_x p_z + \alpha_x \beta p_x mc$$

$$+ \alpha_y \alpha_x p_y p_x + \alpha_y^2 p_y^2 + \alpha_y \alpha_z p_y p_z + \alpha_y \beta p_y mc$$

$$+ \alpha_z \alpha_x p_z p_x + \alpha_z \alpha_y p_z p_y + \alpha_z^2 p_z^2 + \alpha_z \beta p_z mc$$

$$+ \beta \alpha_x mc p_x + \beta \alpha_y mc p_y + \beta \alpha_z mc p_z + \beta^2 m^2 c^2 \tag{6.40}$$

For the right-hand side of Eq. (6.40) to be equal to the left-hand side, the following must be true.

$$\alpha_x^2 = \alpha_y^2 = \alpha_z^2 = \beta^2 = 1 \tag{6.41}$$

$$\alpha_x\alpha_y + \alpha_y\alpha_x = 0$$
$$\alpha_y\alpha_z + \alpha_z\alpha_x = 0 \tag{6.42}$$
$$\alpha_y\alpha_z + \alpha_z\alpha_y = 0$$

$$\alpha_x\beta + \beta\alpha_x = 0$$
$$\alpha_y\beta + \beta\alpha_y = 0 \tag{6.43}$$

If we assume that Eqs. (6.41) – (6.43) are satisfied, then Eq. (6.37) reduces to either

$$W = c\alpha_x p_x + c\alpha_y p_y + c\alpha_z p_z + \beta mc^2 \tag{6.44}$$

or

$$W = -c\alpha_x p_x - c\alpha_y p_y - c\alpha_z p_z - \beta mc^2 \tag{6.45}$$

Making the replacements from Eq. (6.39) in Eq. (6.44), leads to the following form of the Dirac equation.

$$-\frac{\hbar}{ic}\frac{\partial\psi}{\partial t} = \frac{\hbar}{i}\boldsymbol{\alpha}\bullet\nabla\psi + \beta mc\psi \tag{6.46}$$

Dirac showed that Eqs. (6.41) – (6.43) could be satisfied by making α_x, α_y, α_z, and β 4 x 4 matrices. He found that the following so-called Dirac matrices would work.

$$\beta = \begin{bmatrix} 1 & 0 & 0 & 0 \\ 0 & 1 & 0 & 0 \\ 0 & 0 & -1 & 0 \\ 0 & 0 & 0 & -1 \end{bmatrix} \qquad \alpha_x = \begin{bmatrix} 0 & 0 & 0 & 1 \\ 0 & 0 & 1 & 0 \\ 0 & 1 & 0 & 0 \\ 1 & 0 & 0 & 0 \end{bmatrix}$$

$$\tag{6.47}$$

$$\alpha_y = \begin{bmatrix} 0 & 0 & 0 & -i \\ 0 & 0 & i & 0 \\ 0 & -i & 0 & 0 \\ i & 0 & 0 & 0 \end{bmatrix} \qquad \alpha_z = \begin{bmatrix} 0 & 0 & 1 & 0 \\ 0 & 0 & 0 & -1 \\ 1 & 0 & 0 & 0 \\ 0 & -1 & 0 & 0 \end{bmatrix}$$

You can verify that Eqs. (6.41) – (6.43) are satisfied by using the 4 x 4 matrices in Eq. (6.47). For example, it is easy to see that

$$\beta\beta = \alpha_x\alpha_x = \alpha_y\alpha_y = \alpha_z\alpha_z = \begin{bmatrix} 1 & 0 & 0 & 0 \\ 0 & 1 & 0 & 0 \\ 0 & 0 & 1 & 0 \\ 0 & 0 & 0 & 1 \end{bmatrix} \qquad (6.48)$$

The fact that the Dirac matrices are 4 x 4 matrices means that the ψ in Eq. (6.46) is really a 4 x 1 column matrix. That is,

$$\psi(x,y,z,t) = \begin{bmatrix} \psi_1 \\ \psi_2 \\ \psi_3 \\ \psi_4 \end{bmatrix} \qquad (6.49)$$

and there are four values of ψ in the Dirac equation.

The linear approximation of the Hayes field equation is given by Eq. (6.27) as

$$D\psi = K\psi \qquad (6.50)$$

To see how this equation is related to the Dirac equation, we can first use Eq. (6.6) to rewrite Eq. (6.50) as

$$\left(i\frac{\partial}{\partial t} + \nabla \right)\psi = K\psi \qquad (6.51)$$

Multiplying Eq. (6.51) by $\hbar = h/2\pi$ and adding c to the first term to convert from natural units to cgs units, we obtain

$$-\frac{\hbar}{ic}\frac{\partial \psi}{\partial t} = -\hbar \nabla \psi + \hbar K \psi \tag{6.52}$$

From Eq. (6.46), we can write the Dirac equation as

$$-\frac{\hbar}{ic}\frac{\partial \psi}{\partial t} = \frac{\hbar}{i}\boldsymbol{\alpha} \bullet \nabla \psi + \beta mc\psi \tag{6.53}$$

Eqs. (6.52) and (6.53) will be identical if we choose the constant K to be

$$K = \frac{\beta mc}{\hbar} \tag{6.54}$$

and if the quaternian operator ∇ is given by

$$\nabla = i\boldsymbol{\alpha} \bullet \nabla \tag{6.55}$$

The left-hand side of Eq. (6.55) is the quaternion operator

$$\nabla = \mathbf{i}\frac{\partial}{\partial x} + \mathbf{j}\frac{\partial}{\partial y} + \mathbf{k}\frac{\partial}{\partial z} \tag{6.56}$$

where $\mathbf{i}, \mathbf{j}, \mathbf{k}$ are the quaternion unit vectors. As shown in Eq. (A.10) in Appendix A, these unit vectors can be represented by the following three 2 x 2 matrices.

$$\mathbf{i} = \begin{bmatrix} \iota & 0 \\ 0 & -\iota \end{bmatrix} \qquad \mathbf{j} = \begin{bmatrix} 0 & 1 \\ -1 & 0 \end{bmatrix} \qquad \mathbf{k} = \begin{bmatrix} 0 & \iota \\ \iota & 0 \end{bmatrix} \tag{6.57}$$

These can be related to the three Pauli spin matrices given by

$$\sigma_1 = \begin{bmatrix} 0 & 1 \\ 1 & 0 \end{bmatrix} \qquad \sigma_2 = \begin{bmatrix} 0 & -\iota \\ \iota & 0 \end{bmatrix} \qquad \sigma_3 = \begin{bmatrix} 1 & 0 \\ 0 & -1 \end{bmatrix} \tag{6.58}$$

From Eqs. (6.57) and (6.58), note that

$$\mathbf{i} = \iota\sigma_3 \qquad \mathbf{j} = \iota\sigma_2 \qquad \mathbf{k} = \iota\sigma_1 \tag{6.59}$$

from which,

$$\sigma_3 = -\iota \mathbf{i} \qquad \sigma_2 = -\iota \mathbf{j} \qquad \sigma_1 = -\iota \mathbf{k} \qquad (6.60)$$

The 4 x 4 Dirac matrices given by Eq. (6.47) can be written in terms of the 2 x 2 Pauli spin matrices as follows:

$$\alpha_x = \begin{bmatrix} 0 & \sigma_1 \\ \sigma_1 & 0 \end{bmatrix} \qquad \alpha_y = \begin{bmatrix} 0 & \sigma_2 \\ \sigma_2 & 0 \end{bmatrix} \qquad \alpha_z = \begin{bmatrix} 0 & \sigma_3 \\ \sigma_3 & 0 \end{bmatrix} \qquad (6.61)$$

Using Eq. (6.60) in Eq. (6.61), we can write the Dirac matrices as

$$\alpha_x = -\iota \begin{bmatrix} 0 & \mathbf{k} \\ \mathbf{k} & 0 \end{bmatrix} \qquad \alpha_y = -\iota \begin{bmatrix} 0 & \mathbf{j} \\ \mathbf{j} & 0 \end{bmatrix} \qquad \alpha_z = -\iota \begin{bmatrix} 0 & \mathbf{i} \\ \mathbf{i} & 0 \end{bmatrix} \qquad (6.62)$$

Using these values, the right-hand side of Eq. (6.55) can be written as

$$\iota \alpha \bullet \nabla = \begin{bmatrix} 0 & \mathbf{k} \\ \mathbf{k} & 0 \end{bmatrix} \frac{\partial}{\partial x} + \begin{bmatrix} 0 & \mathbf{j} \\ \mathbf{j} & 0 \end{bmatrix} \frac{\partial}{\partial y} + \begin{bmatrix} 0 & \mathbf{i} \\ \mathbf{i} & 0 \end{bmatrix} \frac{\partial}{\partial z} \qquad (6.63)$$

The similarity with the left-hand side given by Eq. (6.56), except for the interchange of the x- and z-axes, is apparent. There are other differences between the quaternion equation, Eq. (6.52), and the Dirac equation, Eq. (6.53). The squares of the quaternion unit vectors are equal to -1, whereas the squares of the Dirac α's are $+1$. But the biggest difference is that the ψ values in the Dirac equation represent probability amplitudes, whereas the ψ value in the Hayes field equation is given by Eq. (6.7) in terms of the electric and magnetic field vectors. Therefore, the entire interpretation of quantum mechanics in terms of probabilities is replaced with an interpretation based on electromagnetic energy densities. The source of spin may be related to the polarizations of the electric and magnetic fields.

6.3 The Conservation Laws

Hayes asserts that the conservation laws are implied by the field equation. From Eqs. (6.15) – (6.18), the expanded field equations are

$$\nabla \cdot \mathbf{B} = 0 \tag{6.64}$$

$$\nabla \cdot \mathbf{E} = \frac{1}{2}\left(B^2 + E^2\right) \tag{6.65}$$

$$\nabla \times \mathbf{B} - \frac{\partial \mathbf{E}}{\partial t} = \mathbf{E} \times \mathbf{B} \tag{6.66}$$

$$\nabla \times \mathbf{E} + \frac{\partial \mathbf{D}}{\partial t} - 0 \tag{6.67}$$

We can rewrite Eqs. (6.67) and (6.66) in the index notation as

$$\varepsilon_{ijk} E_{k,j} + \dot{B}_i = 0 \tag{6.68}$$

and

$$\varepsilon_{ijk} B_{k,j} - \dot{E}_i = \varepsilon_{ijk} E_j B_k \tag{6.69}$$

Multiplying Eq. (6.68) by B_i and Eq. (6.69) by E_i, we obtain

$$\varepsilon_{ijk} E_{k,j} B_i + \dot{B}_i B_i = 0 \tag{6.70}$$

and

$$\varepsilon_{ijk} B_{k,j} E_i - \dot{E}_i E_i = \varepsilon_{ijk} E_j B_k E_i \tag{6.71}$$

Subtracting Eq. (6.71) from Eq. (6.70), we obtain

$$\varepsilon_{ijk}\left(E_{k,j} B_i - B_{k,j} E_i\right) + \dot{B}_i B_i + \dot{E}_i E_i + \varepsilon_{ijk} E_i E_j B_k = 0 \tag{6.72}$$

The last term in Eq. (6.72) is identically zero (interchanging the indexes i and j gives the negative of the term). In the normal Maxwell's equations, this term is the Joule heating term $J_i E_i$. At the fundamental level of the field equation, the ψ-field, before it is organized into particles, is lossless, as there is nothing to heat. Eq. (6.72) can then be rewritten as

$$\varepsilon_{ijk}\left[\left(E_k B_i\right)_{,j} - E_k B_{i,j} + B_{i,j}E_k\right] + \dot{B}_i B_i + \dot{E}_i E_i = 0 \qquad (6.73)$$

or

$$\varepsilon_{ijk}\left(E_j B_k\right)_{,i} + \frac{\partial}{\partial t}\left[\frac{1}{2}\left(B^2 + E^2\right)\right] = 0 \qquad (6.74)$$

which can be rewritten in symbolic notation as

$$\frac{\partial}{\partial t}\left[\frac{1}{2}\left(B^2 + E^2\right)\right] + \nabla \bullet \left(\mathbf{E} \times \mathbf{B}\right) = 0 \qquad (6.75)$$

Eq. (6.75) is the general conservation law. If there are no sources of sinks, so that the divergence of $\mathbf{E} \times \mathbf{B}$ is zero, then $(1/2)\left(B^2 + E^2\right)$ does not vary with time, or is conserved.

If we integrate Eq. (6.75) over a volume V and use the divergence theorem, we can write

$$\frac{\partial}{\partial t}\iiint_V \frac{1}{2}\left(B^2 + E^2\right)dV + \iint_S \left(\mathbf{E} \times \mathbf{B}\right) \bullet d\mathbf{S} = 0 \qquad (6.76)$$

The first term in Eq. (6.76) is the rate of change of the scalar energy-mass-charge, $(1/2)\left(B^2 + E^2\right)$, in a volume. The only way that this value within the volume can change is to transmit or receive the vector energy flux-momentum-electric current, $\mathbf{E} \times \mathbf{B}$, across its boundary. Separate conservation laws for energy, mass, charge, momentum, etc., are special cases of this general conservation law.

6.4 The Uncertainty Principle

Hayes postulates that particles are standing electromagnetic wave packets generated by the nonlinear field equation. In Section 4.6, we saw that a harmonic solution to the wave equation for the electric field vector could be written, from Eq. (4.104), as

$$E_r = A_r e^{i(k\alpha_i x_i \pm \omega t)} = A_r e^{i2\pi\left(\frac{\alpha_i x_i}{\lambda} \pm ft\right)} \qquad (6.77)$$

where

$$\omega = kc = 2\pi f \tag{6.78}$$

$$\lambda = 2\pi/k \tag{6.79}$$

Let $E(x,t)$ be one of the transverse components (say the y-component) of the electric field vector, with amplitude A, propagating in the negative x-direction. Then, from Eq. (6.77), we can write $E(x,t)$ as

$$E(x,t) = Ae^{i(kx+\omega t)} = Ae^{i2\pi\left(\frac{x}{\lambda}+ft\right)} \tag{6.80}$$

Let

$$\zeta = 1/\lambda \tag{6.81}$$

be the spatial frequency, the number of cycles of the wave per unit length. Then we can rewrite Eq. (6.80) as

$$E(x,t) = Ae^{i2\pi(\zeta x+ft)} \tag{6.82}$$

This is a plane wave propagating in the negative x-direction. The wave has a single frequency, f, and wavelength, λ. However, the wave has an unlimited extent in the plus and minus x-direction. To confine the wave to a small region along the x-axis, we need to add up waves of slightly different frequencies to form a so-called *wave packet*. We can therefore write $E(x,t)$ in the form

$$E(x,t) = \int_{-\infty}^{\infty} A(f)e^{i2\pi(\zeta x+ft)}df \tag{6.83}$$

where $A(f)$ is assumed to be peaked about some central frequency, f_0.

We could have taken the spatial frequency ζ to be the independent variable, in which case

$$E(x,t) = \int\limits_{-\infty}^{\infty} A(\zeta)e^{i2\pi(\zeta x+ft)}d\zeta \tag{6.84}$$

Eqs. (6.83) and (6.84) look like inverse Fourier transforms of the temporal or spatial frequency spectra. Indeed, at time $t=0$, Eq. (6.84) reduces to

$$E(x,0) = \int\limits_{-\infty}^{\infty} A(\zeta)e^{i2\pi\zeta x}d\zeta \tag{6.85}$$

which is just the inverse Fourier transform of $A(\zeta)$. The Fourier transform, $A(\zeta)$, of the spatial function, $E(x,0)$, is given by

$$A(\zeta) = \int\limits_{-\infty}^{\infty} E(x,0)e^{-i2\pi\zeta x}dx \tag{6.86}$$

Similarly, at $x=0$, Eq. (6.83) reduces to

$$E(0,t) = \int\limits_{-\infty}^{\infty} A(f)e^{i2\pi ft}df \tag{6.87}$$

which is just the inverse Fourier transform of $A(f)$. The Fourier transform, $A(f)$, of the temporal function, $E(0,t)$, is given by

$$A(f) = \int\limits_{-\infty}^{\infty} E(0,t)e^{-i2\pi\zeta x}dt \tag{6.88}$$

An example of a wave packet, $E(x,0)$, is shown in Fig. 6.2. This function is formed by multiplying a cosine wave with spatial frequency ζ_0 by the Gaussian function $g(x/2)$ shown in Fig. 6.3. The Fourier transform of the wave packet $E(x,0)$ in Fig. 6.2 will be given by Eq. (6.86). It will be the Fourier transform of the Gaussian function shown in Fig. 6.3, centered at the spatial frequencies $\pm\zeta_0$.

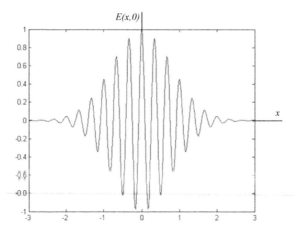

Fig. 6.2 A wave packet with Gaussian envelope

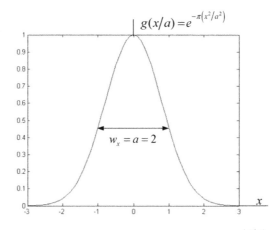

Fig. 6.3 The Gaussian function $g(x/a) = e^{-\pi\left(x^2/a^2\right)}$

The Fourier transform of the Gaussian function $g(t) = e^{-\pi t^2}$ is calculated in Appendix B. It is equal to $G(f) = e^{-\pi f^2}$. That is, the Fourier transform of this Gaussian function is the same Gaussian function. A plot of $g(t)$ is shown in Fig. 6.4. We define the width of $g(t)$ to be $w_g = 1$ as shown in Fig. 6.4.

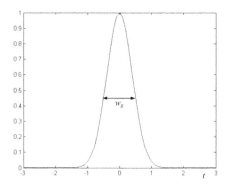

Fig. 6.4 The Gaussian function, $g(t)$

As shown in Appendix B, we can indicate the Fourier transform pairs for a Gaussian function using the notation

$$e^{-\pi t^2} \Leftrightarrow e^{-\pi f^2} \tag{6.89}$$

From the Fourier transform scaling theorem, we can write

$$g(t/a) \Leftrightarrow |a| G(af) \tag{6.90}$$

This means that the Fourier transform of a Gaussian function of width a will be

$$e^{-\pi(t^2/a^2)} \Leftrightarrow |a| e^{-\pi a^2 f^2} \tag{6.91}$$

Therefore, the Fourier transform of the spatial signal $e^{-\pi(x^2/a^2)}$ shown in Fig. 6.3 will be $|a| e^{-\pi a^2 \zeta^2}$. This Fourier transform for a value of $a = 2$ is shown in Fig. 6.5.

In Fig. 6.3, the width of the Gaussian function is $w_x = a$ and in Fig. 6.5, the width of its Fourier transform is $w_\zeta = 1/a$. The product of these two widths is

$$w_x w_\zeta = 1 \tag{6.92}$$

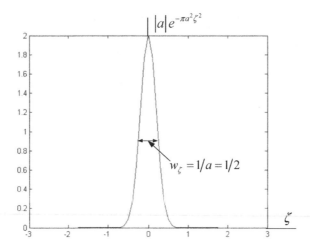

Fig. 6.5 The Fourier transform of the Gaussian
function in Fig. 6.3

If we think of the wave packet as representing the location of a particle along the x-axis, then we can think of its width, w_x, as the uncertainty, Δx, of its location. That is,

$$\Delta x = w_x \qquad (6.93)$$

Similarly, our uncertainty, $\Delta \zeta$, of the spatial frequency of the waves in the wave packet, can be taken to be the width of the Fourier transform in Fig. 6.5. That is,

$$\Delta \zeta = w_\zeta \qquad (6.94)$$

Substituting Eqs. (6.93) and (6.94) in Eq. (6.92), we obtain

$$\Delta x \Delta \zeta = 1 \qquad (6.95)$$

All of these results apply if we consider time signals and temporal frequencies. If you stand at $x = 0$, and watch the wave packet in Fig. 6.2 go by, then you will see the same wave packet as in Fig. 6.2, but now a function of time. The Gaussian function in Fig. 6.3 will be a function of time, and the corresponding Fourier transform in Fig. 6.5 will be a function of the temporal frequency, f.

The widths of these Gaussian functions will be $w_t = a$ and $w_f = 1/a$, so that

$$w_t w_f = 1 \tag{6.96}$$

Our uncertainty as to when the wave packet goes by will be

$$\Delta t = w_t \tag{6.97}$$

and our uncertainty as to the frequency content of the wave packet will be

$$\Delta f = w_f \tag{6.98}$$

Substituting Eqs. (6.97) and (6.98) in Eq. (6.96), we obtain

$$\Delta t \Delta f = 1 \tag{6.99}$$

The uncertainty relations given by Eqs. (6.95) and (6.99) have nothing to do with physics. They are inherent relationships between a function and its Fourier transform. As a signal gets narrower and narrower, its Fourier transform spreads out, and vice versa.

The uncertainty principle in physics comes in when we associate energy with frequency, as Einstein did when explaining the photoelectric effect in 1905, where the energy associated with a photon of light is given by

$$E = hf \tag{6.100}$$

where h is Plank's constant, and when we associate momentum with wavelength, as de Broglie did in 1924, when he proposed that particles with momentum p have a wavelength given by

$$\lambda = h/p \tag{6.101}$$

Using Eq. (6.81), we can rewrite Eq. (6.101) in terms of the spatial frequency as

$$p = h/\lambda = h\zeta \tag{6.102}$$

From Eqs. (6.100) and (6.102), we can write

$$\Delta E = h\Delta f \tag{6.103}$$

and

$$\Delta p = h\Delta \zeta \tag{6.104}$$

Substituting Eq. (6.104) in Eq. (6.95), that uncertainty relation becomes

$$\Delta x \Delta p = h \tag{6.105}$$

Similarly, substituting Eq. (6.103) in Eq. (6.99), that uncertainty relation becomes

$$\Delta t \Delta E = h \tag{6.106}$$

Eqs. (105) and (106) form the basis of what is known as the Heisenberg uncertainty principle. It is often discussed in connection with the limits on experimental measurements and is sometimes expressed as

$$\Delta x \Delta p \geq (\sim h) \tag{6.107}$$

and

$$\Delta t \Delta E \geq (\sim h) \tag{6.108}$$

Much has been written about the meaning of the uncertainty principle and how it may imply that the universe is non-causal. Hayes rejects this interpretation and rather interprets it as a granularity principle.

6.5 The Relativistic Invariance of the Field Equation

For the field equation to be correct and to represent physical reality, it must be the same in all inertial reference frames. The whole point of special relativity is that there is no preferred reference frame in which the equations of physics are true.

From Eqs. (6.5) – (6.7), the field equation is given be

$$D\psi = \frac{1}{2}\psi * \iota\psi \qquad (6.109)$$

where

$$D = \iota\frac{\partial}{\partial t} + \nabla = \iota\frac{\partial}{\partial t} + \mathbf{i}\frac{\partial}{\partial x} + \mathbf{j}\frac{\partial}{\partial y} + \mathbf{k}\frac{\partial}{\partial z} \qquad (6.110)$$

and

$$\psi = \mathbf{B} + \iota\mathbf{E} \qquad (6.111)$$

Assume that the values of (x, y, z, t) in this equation are with respect to some stationery reference frame S. In a reference frame S' moving uniformly with respect to S with a velocity $\mathbf{u} = u\boldsymbol{\alpha}$, the form of the field equation must be

$$D'\psi' = \frac{1}{2}\psi *' \iota'\psi' \qquad (6.112)$$

In this section, we will show that Eq. (6.112) is true.

We know that Maxwell's equations are relativistically invariant. After all, we derived them from special relativity and Coulomb's law and it was the absence of a preferred reference frame for electrodynamics that led Einstein to his theory of relativity in the first place.

In Section 4.7, Eqs. (4.116) – (4.120), we showed that Maxwell's equations can be written as a single quaternion equation in the form

$$D\psi = C \qquad (6.113)$$

where

$$D = \left(\frac{i}{c}\frac{\partial}{\partial t} + \nabla\right) \qquad (6.114)$$

$$\psi = \mathbf{B} + i\mathbf{E} \qquad (6.115)$$

and

$$C = -i\rho + \frac{1}{c}\mathbf{J} \qquad (6.116)$$

For simplicity, we will assume units in which $c = 1$, and rewrite Eqs. (6.114) and (6.116) as

$$D = i\frac{\partial}{\partial t} + \mathbf{V} \tag{6.117}$$

$$C = -i\rho + \mathbf{J} \tag{6.118}$$

We should note that the value of ψ could have been written with a minus sign instead of a plus sign as

$$\bar{\psi} = \mathbf{B} - i\mathbf{E} \tag{6.119}$$

This, in fact, is the convention used by L. Silberstein in his book *The Theory of Relativity*, published in 1914. Hayes, on the other hand, uses the plus sign as in Eq. (6.115). If you follow the same derivation given in Section 4.7, this time subtracting the equations instead of adding them, and assuming $c = 1$, you will see that Maxwell's equations can be written as the single quaternion equation

$$\overline{D}\bar{\psi} = \overline{C} \tag{6.120}$$

where

$$\overline{D} = -i\frac{\partial}{\partial t} + \mathbf{V} \tag{6.121}$$

$$\overline{C} = i\rho + \mathbf{J} \tag{6.122}$$

In Section 2.10, we saw in Eqs. (2.44) – (2.47) that the Lorentz transformation is given by

$$\mathbf{r}' = \mathbf{r} + (\gamma - 1)(\boldsymbol{\alpha} \bullet \mathbf{r})\boldsymbol{\alpha} + i\beta\gamma l\boldsymbol{\alpha}$$
$$l' = \gamma(l - i\beta\boldsymbol{\alpha} \bullet \mathbf{r}) \tag{6.123}$$

where

$$\beta = \frac{u}{c} \tag{6.124}$$

$$\gamma = \frac{1}{\sqrt{1 - \beta^2}} \tag{6.125}$$

$$l = ict \tag{6.126}$$

We showed in Section 2.10 that the two Lorentz transformation equations in Eq. (6.123) can be written as the single quaternion equation

$$q' = QqQ \tag{6.127}$$

where

$$q = (l, \mathbf{r}) = l + \mathbf{r} \tag{6.128}$$

$$q' = (l', \mathbf{r}') = l' + \mathbf{r}' \tag{6.129}$$

$$Q = \frac{1}{\sqrt{2}}\left(\sqrt{1+\gamma} + \mathbf{a}\sqrt{1-\gamma}\right)$$
$$= \cos\varphi + \mathbf{a}\sin\varphi \tag{6.130}$$

where

$$2\varphi = \tan^{-1}(i\beta) = \tan^{-1}\left(i\frac{u}{c}\right) \tag{6.131}$$

Silberstein calls a quaternion of the form given in Eq. (6.128), namely, $q = l + \mathbf{r} = ict + \mathbf{r}$, a physical quaternion. Such a physical quaternion will transform to the S' frame of reference according to Eq. (6.127).

From Eq. (6.126), setting $c = 1$, we can write

$$\frac{\partial}{\partial t} = \frac{\partial}{\partial l}\frac{\partial l}{\partial t} = i\frac{\partial}{\partial l} \tag{6.132}$$

Substituting Eq. (6.132) into Eq. (6.121), we can write $\overline{D}$ as

$$\overline{D} = \frac{\partial}{\partial l} + \nabla \tag{6.133}$$

Comparing the components of $\overline{D}$ in Eq. (6.133) with those of q in Eq. (6.128), we see that $\overline{D}$ will transform the same as q, namely, from Eq. (6.127)

$$\overline{D}' = Q\overline{D}Q \tag{6.134}$$

Recall from Appendix A that the complex conjugate of $q = l + \mathbf{r}$ in Eq. (6.128) is $q_c = l - \mathbf{r}$ and the complex conjugate of $Q = \cos\varphi + \boldsymbol{\alpha}\sin\varphi$ in Eq. (6.130) is $Q_c = \cos\varphi - \boldsymbol{\alpha}\sin\varphi$. Inasmuch as changing the signs of $\mathbf{r}$ and $\boldsymbol{\alpha}$ leaves the Lorentz transformation in Eq. (6.123) unchanged, then q_c will transform as

$$q_c' = Q_c q_c Q_c \qquad (6.135)$$

How does D in Eq. (6.117) transform? Using Eq. (6.132), we can write Eq. (6.117) as

$$D = -\frac{\partial}{\partial l} + \nabla = -\left(\frac{\partial}{\partial l} - \nabla\right) \qquad (6.136)$$

or, using Eq. (6.133)

$$D = -\bar{D}_c \qquad (6.137)$$

It also follows that

$$D_c = -\bar{D} \qquad (6.138)$$

Substituting Eq. (6.138) into Eq. (6.134), we see that D_c transforms as

$$-D_c' = Q(-D_c)Q$$

or

$$D_c' = QD_cQ \qquad (6.139)$$

Recall from Appendix A that the conjugate of a quaternion product is the product of the conjugates in reverse order. Therefore, from Eq. (6.139), we can immediately write

$$D' = Q_c D Q_c \qquad (6.140)$$

In a similar way, we see that $\bar{C}$ in Eq. (6.122) will transform as

$$\bar{C}' = Q\bar{C}Q \tag{6.141}$$

and $C = -\bar{C}_c$ in Eq. (6.118) will transform as

$$C' = Q_c C Q_c \tag{6.142}$$

Now in frame S, the single Maxwell quaternion equation from Eq. (6.113) is

$$D\psi = C \tag{6.143}$$

and in frame S' it will be written as

$$D'\psi' = C' \tag{6.144}$$

We know that D transforms according to Eq. (6.140) and C transforms according to Eq. (6.142). How does ψ transform?

If we substitute Eqs. (6.140) and (6.142) in Eq. (6.144), we obtain

$$Q_c D Q_c \psi' = Q_c C Q_c$$

from which

$$D Q_c \psi' = C Q_c \tag{6.145}$$

Using the value of C from Eq. (6.143) in Eq. (6.145), we obtain

$$D Q_c \psi' = D \psi Q_c$$

from which

$$Q_c \psi' = \psi Q_c \tag{6.146}$$

Premultiplying both sides of Eq. (6.146) by Q, we obtain

$$\psi' = Q\psi Q_c \tag{6.147}$$

In frame S, the alternate version of the single Maxwell quaternion equation given in Eq. (6.120) is

$$\bar{D}\bar{\psi} = \bar{C} \tag{6.148}$$

and in frame S' it will be written as

$$\bar{D}'\bar{\psi}' = \bar{C}'$$

(6.149)

We know that $\bar{D}$ transforms according to Eq. (6.134) and $\bar{C}$ transforms according to Eq. (6.141).

To find out how $\bar{\psi}$ transforms, substitute Eqs. (6.134) and (6.141) in Eq. (6.149) to obtain

$$Q\bar{D}Q\bar{\psi}' = Q\bar{C}Q$$

(6.150)

from which

$$\bar{D}Q\bar{\psi}' = \bar{C}Q$$

(6.151)

Using the value of $\bar{C}$ from Eq. (6.148) in Eq. (6.151), we obtain

$$\bar{D}Q\bar{\psi}' = \bar{D}\bar{\psi}Q$$

(6.152)

from which

$$Q\bar{\psi}' = \bar{\psi}Q$$

(6.153)

Premultiplying both sides of Eq. (6.153) by Q_c, we obtain

$$\bar{\psi}' = Q_c\bar{\psi}Q$$

(6.154)

Summarizing the results from Eqs. (6.140), (6.139), (6.147), and (6.154), we have the following transformation rules.

$$D' = Q_c D Q_c$$

(6.155)

$$D_c' = Q D_c Q$$

(6.156)

$$\psi' = Q\psi Q_c$$

(6.157)

$$\bar{\psi}' = Q_c\bar{\psi}Q$$

(6.158)

We now return to the field equation

$$D\psi = \frac{1}{2}\psi * \iota \psi$$

(6.159)

where

$$\psi = \mathbf{B} + \iota \mathbf{E} \tag{6.160}$$

is the same as Ψ in Eq. (6.115). In Eq. (6.159)

$$\psi^* = \mathbf{B} - \iota \mathbf{E} \tag{6.161}$$

which is the same as $\overline{\Psi}$ in Eq. (6.119). Therefore, from Eqs. (6.157) and (6.158), we see that ψ and ψ^* transform as

$$\psi' = Q\psi Q_c \tag{6.162}$$
$$\psi^{*'} = Q_c \psi^* Q \tag{6.163}$$

It is important to note that ψ^* is not a quaternion conjugate as defined in Appendix A where the vector part of the quaternion is negated. Rather it is a kind of complex conjugate of a complex quaternion.

Another important point is that the unit imaginary scalar ι in Eq. (6.159) is a special case of $q = \iota a + \mathbf{b}$ or $q_c = \iota a - \mathbf{b}$, where $a = 1$ and $\mathbf{b} = 0$. It transforms as

$$\iota' = Q\iota Q \tag{6.164}$$

or

$$\iota' = Q_c \iota Q_c \tag{6.165}$$

whichever is more convenient.

We are now in a position to verify Eq. (6.112). If we pre-multiply and post-multiply both sides of Eq. (6.159) by Q_c and insert $QQ_c = Q_c Q = 1$, we can write

$$Q_c D Q_c Q \psi Q_c = \frac{1}{2} Q_c \psi^* Q Q_c \iota Q_c Q \psi Q_c \tag{6.166}$$

Using the transformations from Eqs. (6.155), (6.162), (6.163), and (6.165), we obtain

$$D'\psi' = \frac{1}{2} \psi^* \iota' \psi' \tag{6.167}$$

which verifies Eq. (6.112). Note that the position of ι in Eq. (6.159) is important for the field equation to be invariant under a Lorentz transformation as can be seen in Eq. (6.166).

Hayes summarizes the importance of the relativistic invariance of the field equation when he writes the following on page 21 of his book.

> "The relativistic invariance of the field equation is an extremely convincing piece of evidence that the unified field theory is correct. For, with the possible exception of one or two equations like the Klein-Gordon equation which have not been too successful in explaining physical phenomena, the only relativistically invariant equations are the field equation $D\psi = (1/2)\psi * \iota\psi$, Dirac's equation $D\psi = K\psi$, and Maxwell's equation $D\psi = C$; and the last two are approximations of the first. The burden therefore rests upon the skeptic to produce an alternative invariant equation that accounts for Dirac's and Maxwell's equations, or disprove relativity; or accept the field equation."

6.6 The Gravitational Field

According to Hayes, gravitational fields are ψ-fields, i.e., electromagnetic fields, between neutral particles. Neutral particles will be ψ-fields with equal amounts of right-handed and left-handed energy, or equals amount of positive and negative charge – as with an electron-proton pair. The term $\frac{1}{2}\left(B^2 + E^2\right)$ in Eq. (6.16) represents all of matter, charge-mass-energy. The term $\mathbf{E} \times \mathbf{B}$ in Eq. (6.17) is defined to be momentum. The constant approximation to the field equation is Maxwell's equations, which now include the gravito-electromagnetic equations derived from Newton's law of gravitation and special relativity in Chapter 5. These gravito-electromagnetic fields are just the same electric and magnetic fields of electromagnetics.

From Eq. (4.90), we can write the four Maxwell equations in the index notation, with $c = 1$, as

$$\varepsilon_{ijk}E_{k,j} = -\dot{B}_i \tag{6.168}$$

$$E_{i,i} = \rho \tag{6.169}$$

$$\varepsilon_{ijk} B_{k,j} = J_i + \dot{E}_i \tag{6.170}$$

$$B_{i,i} = 0 \tag{6.171}$$

Taking the cross product of Eq. (6.168) with E_s, we can write

$$\varepsilon_{ris} \varepsilon_{ijk} E_{k,j} E_s = -\varepsilon_{ris} \dot{B}_i E_s \tag{6.172}$$

Taking the cross product of Eq. (6.170) with B_s, we can write

$$\varepsilon_{ris} \varepsilon_{ijk} B_{k,j} B_s = \varepsilon_{ris} J_i B_s + \varepsilon_{ris} \dot{E}_i B_s \tag{6.173}$$

Adding Eqs. (6.172) and Eq. (6.173), we obtain

$$-\varepsilon_{irs} \varepsilon_{ijk} \left(E_{k,j} E_s + B_{k,j} B_s \right) = \varepsilon_{ris} J_i B_s + \varepsilon_{ris} \left(\frac{\partial E_i}{\partial t} B_s + \frac{\partial B_s}{\partial t} E_i \right)$$

which leads to

$$-\left(\delta_{rj} \delta_{sk} - \delta_{rk} \delta_{sj} \right) \left(E_{k,j} E_s + B_{k,j} B_s \right) = \varepsilon_{ris} J_i B_s + \varepsilon_{ris} \left[\frac{\partial}{\partial t} \left(E_i B_s \right) \right]$$

or

$$-E_{s,r} E_s + E_{r,s} E_s - B_{s,r} B_s + B_{r,s} B_s$$

$$= \varepsilon_{ris} J_i B_s + \left[\frac{\partial}{\partial t} \left(\varepsilon_{ris} E_i B_s \right) \right] \tag{6.174}$$

Now look at the first two terms of Eq. (6.174). These two terms can be written as

$$E_{r,s} E_s - E_{s,r} E_s = \left(E_r E_s \right)_{,s} - E_r E_{s,s} - \left(\frac{1}{2} E_s E_s \right)_{,r} \tag{6.175}$$

Substituting Eq. (6.169) in Eq. (6.175), we obtain

$$
\begin{aligned}
E_{r,s}E_s - E_{s,r}E_s &= \left(E_r E_s \right)_{,s} - \rho E_r - \left(\frac{1}{2} E_j E_j \delta_{sr} \right)_{,s} \\
&= \left(E_r E_s - \frac{1}{2} E_j E_j \delta_{rs} \right)_{,s} - \rho E_r
\end{aligned}
\tag{6.176}
$$

We can write Eq. (6.176) as

$$
E_{r,s}E_s - E_{s,r}E_s = T^E_{rs,s} - \rho E_r
\tag{6.177}
$$

where

$$
T^E_{rs} = E_r E_s - \frac{1}{2} E_j E_j \delta_{rs}
\tag{6.178}
$$

is called the *electric stress tensor*. The matrix of T^E_{rs} is thus

$$
T^E_{rs} =
\begin{bmatrix}
E_1 E_1 - \dfrac{1}{2} E_j E_j & E_1 E_2 & E_1 E_3 \\
E_2 E_1 & E_2 E_2 - \dfrac{1}{2} E_j E_j & E_2 E_3 \\
E_3 E_1 & E_3 E_2 & E_3 E_3 - \dfrac{1}{2} E_j E_j
\end{bmatrix}
\tag{6.179}
$$

The third and fourth terms of Eq. (6.174) are

$$
B_{r,s}B_s - B_{s,r}B_s = \left(B_r B_s \right)_{,s} - B_r B_{s,s} - \left(\frac{1}{2} B_s B_s \right)_{,r}
\tag{6.180}
$$

which, using Eq. (6.171), can be written as

$$
B_{r,s}B_s - B_{s,r}B_s = \left(B_r B_s - \frac{1}{2} B_j B_j \delta_{rs} \right)_{,s}
\tag{6.181}
$$

We can write Eq. (6.181) as

$$
B_{r,s}B_s - B_{s,r}B_s = T^M_{rs,s}
\tag{6.182}
$$

where

$$T_{rs}^M = B_r B_s - \frac{1}{2} B_j B_j \delta_{rs} \tag{6.183}$$

is called the *magnetic stress tensor*. The matrix of T_{rs}^M is thus

$$T_{rs}^M = \begin{bmatrix} B_1 B_1 - \frac{1}{2} B_j B_j & B_1 B_2 & B_1 B_3 \\[2ex] B_2 B_1 & B_2 B_2 - \frac{1}{2} B_j B_j & B_2 B_3 \\[2ex] B_3 B_1 & B_3 B_2 & B_3 B_3 - \frac{1}{2} B_j B_j \end{bmatrix} \tag{6.184}$$

The total electromagnetic stress tensor, T_{rs}, is defined as

$$T_{rs} = T_{rs}^E + T_{rs}^M \tag{6.185}$$

Substituting Eqs. (6.177), (6.182), and (6.185) into Eq. (6.174), we obtain

$$T_{rs,s} - \rho E_r = \varepsilon_{ris} J_i B_s + \left[\frac{\partial}{\partial t} \left(\varepsilon_{ris} E_i B_s \right) \right]$$

or

$$\rho E_r + \varepsilon_{ris} J_i B_s = T_{rs,s} - \frac{\partial}{\partial t} \left(\varepsilon_{ris} E_i B_s \right) \tag{6.186}$$

Eq. (6.186) gives the forces per unit volume on the charges and currents in terms of the field quantities alone. If we integrate Eq. (6.186) over the volume V bounded by the closed surface S and use the divergence theorem, we obtain

$$\int_V \rho E_r dV + \int_V \varepsilon_{ris} J_i B_s dV = \int_S T_{rs} n_s dS - \frac{\partial}{\partial t} \int \varepsilon_{ris} E_i B_s dV \tag{6.186}$$

Hayes makes two interpretations of Eq. (6.186). He first postulates his field equation and its constant approximation,

Maxwell's equations, defines $\mathbf{E} \times \mathbf{B}$, or $\varepsilon_{ris} E_i B_s$, as momentum, and defines the left-hand side of Eq. (6.186) to be what is meant by force. Then the left-hand side of Eq. (6.186) is the total force, K_r on all of the classical particles in V exerted by the rest of the field in V. Assuming that the field vanishes on the bounding surface, S, Eq. (6.186) reduces to

$$K_r = -\frac{\partial}{\partial t} \int \varepsilon_{ris} E_i B_s dV \qquad (6.187)$$

which is Newton's law of motion. In Hayes' theory, classical particles are intense regions of the field surounded by fictitious spherical control surfaces. Eq. (6.187) states that the force K_r on these classical particles is equal and opposite to the force on the rest of the field in V and is equal to the time rate of change of momentum of the field. Hayes extends the argument to collections of classical particles and concludes,

> Given Newton's law of motion, the rest of classical mechanics follows at once by purely mathematical deduction, including Lagrange's equations, Hamilton's equations, Hamilton's principle, and so forth.

The assumption that the surface integral in Eq. (6.186) vanishes because the field values fall off inversely as the square of the distance in probably not justified. In fact, in his discussion of the gravitational field, Hayes suspects that this surface integral represents the gravitational force. He makes the following argument.

If you surround the earth with a surface between it and the sun, the field ψ could not forever be zero everywhere on this surface, otherwise the sun could have no knowledge of the existence of the earth and vice versa. No gravitational force would exist. It is the non-zero field on the bounding surface that transmits the gravitational force.

Suppose all of the positive charges of the earth were concentrated at its center. They would produce an **E**-field (and perhaps a **B**-field), which, just from geometry, would drop off inversely as the square of the distance to the surface S. This would produce a force at S given by the surface integral in Eq. (6.186). The positive charges on the sun would produce a similar force on S. The same argument can be made if you concentrate all of the negative charges on the center of the earth (and sun). But the force given by the surface integral in Eq. (6.186) is quadratic in **E** and **B**, so the forces from positive and negative particles add. In fact, the surface integral in Eq. (6.186) is proportional to $\frac{1}{2}\left(B^2 + E^2\right)$, which in the natural units of the field equation is what we mean by mass. Therefore, the surface integral in Eq. (6.186) suggests a positive force that falls off as the square of the distance and is proportional to the masses of the attracting bodies.

This view of gravity is, of course, quite different from the prevailing conventional wisdom of Einstein's general relativity and curved space. Hayes comments on this as follows:

> J. A. Wheeler, in his article, with S. Tilson, "Dynamics of Space-Time", in the Dec. '63 issue of International Science and Technology, points out that Einstein's theory is completely summarized in the statement:
> *Intrinsic curvature − extrinsic curvature = 16π(energy density)*
> The field theory embodied in $D \psi = (1/2)\psi * \iota \psi$, is summarized in the statement:
> *Quaternionic rate of change of the field*
> $$= energy\ density\ quaternion$$
> Are the two theories saying the same thing in different mathematics or are they fundamentally different theories?

Einstein's field equation can be written (in units where the speed of light and the gravitational constant are set to 1) as

$$R_{\mu v} - \frac{1}{2} R g_{\mu v} = 8\pi T_{\mu v} \qquad (6.188)$$

where $R_{\mu v}$, is the Ricci tensor (a contraction of the Riemann curvature tensor), R is the curvature scalar (a contraction of the Ricci tensor), $g_{\mu v}$ is the metric tensor, and $T_{\mu v}$ is the energy-momentum tensor.

There are significant differences between the ideas of Einstein's field equation, Eq. (6.188), and Hayes' field equation, Eq. (6.159). The energy-momentum tensor, $T_{\mu\nu}$, on the right-hand side of Eq. (6.188) represents the sources of matter and energy which causes the curvature of space-time described by the left-hand side of Eq. (6.188). This energy-momentum tensor may contain terms in the electromagnetic stress tensor of Eq. (6.185), but also may contain momentum flow terms associated with physical particles. Both field equations are nonlinear; however, the nonlinearities in the Hayes equation occur only on the right-hand side, while the left-hand side of Eq. (6.188) is a complicated nonlinear function of the metric tensor. The curvature tensor, from which the Ricci tensor and curvature scalar are derived, contains both derivatives and products of Christoffel symbols, which contain products of the metric tensor with its derivative. Most solutions of Eq. (6.188) are greatly simplified, often linearized, solutions. The solutions of Eq. (6.188) represent curved space-time, and point particles then move along the geodesics of this curved space-time. By contrast, the Hayes field equation is a nonlinear partial differential equation for the regular electric and magnetic field vectors in normal flat space. The one similarity between the two field equations is that after symmetry and constraints are taken into account, Eq. (6.188), reduces to six truly independent equations, and the Hayes field equation solves for the six components of the electric and magnetic fields. The major difference, of course, is that the Einstein field equation is essentially a theory of gravity, while the Hayes field equation purports not only to explain gravity, but also to have as approximations Dirac's equation of relativistic quantum mechanics and Maxwell's equations of electrodynamics.

6.7 Natural System of Units

The Hayes field equation, given from Eq. (6.159) as

$$\mathrm{D}\psi = \frac{1}{2}\psi * \iota\psi \tag{6.189}$$

is given in natural units with no arbitrary constants that need to be tweaked. If one were to get solutions to the eight coupled, nonlinear partial differential equations given by Eqs. (6.19)-(6.26), what would they mean? How would the numbers relate to physical units in the SI or cgs system of units?

The value of the speed of light, c, has already been set to 1, which defines a new fundamental unit of time. In his book, Hayes suggests that setting the ratio of charge to mass of the electron to 1 will define a new fundamental unit of mass. The third fundamental unit of length could then be determined by comparing solutions of the field equation to experiments, or, as Hayes speculates, perhaps the value of Plank's constant, or the reduced Planck's constant is 1.

In my novel, *Peggy's Discovery*, Peggy and her uncle conclude that the field equation requires the speed of light $c=1$, the reduced Planck's constant $\hbar = h/2\pi = 1$, and the gravitational constant $G=1$. Hayes never suggested that the value of the gravitational constant should be set to 1, but then he did not know about Peggy's derivation of the gravito-electromagnetic equations in Chapter 5. Setting $G=1$ in that derivation and equating mass with charge as the field equation asserts, makes the derivation identical to that with Coulomb's law. This means there is only one set of Maxwell's equations, corresponding to the constant approximation of the field equation, which includes both electromagnetic and gravitational fields.

The values of the three fundamental constants, in SI units, are

$$c = 3 \times 10^8 \text{ m s}^{-1} \tag{6.190}$$
$$\hbar = h/2\pi = 1.0546 \times 10^{-34} \text{ kg m}^2 \text{ s}^{-1} \tag{6.191}$$
$$G = 6.674 \times 10^{-11} \text{ kg}^{-1} \text{ m}^3 \text{ s}^{-2} \tag{6.192}$$

In terms of the fundamental units of time, T, length, L, and mass, M, the units of the three fundamental constants are

$$[c] = \text{L T}^{-1} \tag{6.193}$$
$$[\hbar] = \text{M L}^2 \text{ T}^{-1} \tag{6.194}$$
$$[G] = \text{M}^{-1} \text{ L}^3 \text{ T}^{-2} \tag{6.195}$$

Using the values in Eqs. (6.190) – (6.192) in Eqs. (6.193) – (6.195), we can solve for the values of T, L, and M is SI units. These will be the fundamental units of time, length, and mass in the field equation in Eq. (6.189). For example, if a solution to Eq. (6.189) produces a particular value of time, the corresponding time in seconds will be that value times T.

Multiplying Eq. (6.194) by Eq. (6.195), we can write

$$\hbar G = L^5\, T^{-3} \tag{6.196}$$

Substituting T from Eq. (6.193) in Eq. (6.196), we obtain

$$\hbar G = c^3 L^2$$

from which

$$L = \sqrt{\frac{\hbar G}{c^3}} \tag{6.197}$$

Substituting the values in Eqs. (6.190) – (6.192) in Eq. (6.197), we see that the fundamental unit of length, L, is

$$L = 1.616 \times 10^{-35}\ \text{m} \tag{6.198}$$

and is called the *Planck length.*

Substituting L from Eq. (6.193) in Eq. (6.196), we obtain

$$\hbar G = c^5 T^2$$

from which

$$T = \sqrt{\frac{\hbar G}{c^5}} \tag{6.199}$$

Substituting the values in Eqs. (6.190) – (6.192) in Eq. (6.199), we see that the fundamental unit of time, T, is

$$T = 5.391 \times 10^{-44}\ \text{s} \tag{6.200}$$

and is called the *Planck time.* It is equal to the time for light to travel a distance equal to the Planck length.

Finally, to find the fundamental unit of mass, M, divide Eq. (6.194) by Eq. (6.193) to obtain

$$M = \frac{\hbar}{c} L^{-1} \tag{6.201}$$

Substituting Eq. (6.197) in Eq. (6.201), we obtain

$$M = \sqrt{\frac{\hbar c}{G}} \tag{6.202}$$

Substituting the values in Eqs. (6.190) – (6.192) in Eq. (6.202), we see that the fundamental unit of mass, M, is

$$M = 2.177 \times 10^{-8} \text{ kg} \tag{6.203}$$

and is called the *Planck mass*. Using $E = mc^2$, this mass corresponds to an energy of 1.96×10^9 J, or 1.96×10^{16} ergs.

Hayes suggests that it may be possible to experimentally verify the field equation in Eq. (6.189). He writes

> An intense modulated radio or light beam should produce transformer action by the revised Ampere's Law. An intense, coherent light beam, split and mirrored to pass through a region in opposite directions, should produce a zero $\mathbf{E} \times \mathbf{H}$ but a non-zero $1/2(H^2 + E^2)$ and so according to the revised Gauss' Law should produce an electrostatic field.

In my novel, *Peggy's Discovery*, Peggy and her friends devise and carry out just such an experiment. This is easy to do in a novel. To learn what the experiment is and how it turns out, you will need to read the novel!

Appendix A

Introduction to Quaternions

A.1 Definition of a Quaternion

A quaternion is made up of a *scalar* part, s, and a *vector* part, $\mathbf{r}$. We can write the quaternion q as

$$q = (s, \mathbf{r}) \tag{A.1}$$

where

$$\mathbf{r} = x\mathbf{i} + y\mathbf{j} + z\mathbf{k} \tag{A.2}$$

Quaternions are often written as the sum of the scalar and vector parts. Thus, we can write

$$q = s + \mathbf{r} = s + x\mathbf{i} + y\mathbf{j} + z\mathbf{k} \tag{A.3}$$

In Eq. (A.3) s, x, y, and z are scalars and $\mathbf{i}, \mathbf{j}, \mathbf{k} = \sqrt{-1}$. It follows that

$$\mathbf{ii} = \mathbf{jj} = \mathbf{kk} = -1 \tag{A.4}$$

As shown in Eq. (A.2) and Fig. A.1, $\mathbf{i}$, $\mathbf{j}$, and $\mathbf{k}$ are the ordinary unit vectors of 3-dimensional space. However, they are not exactly the same, because Eq. (A.4) shows that the product of these unit vectors with themselves give -1 rather than $+1$. The result of multiplying one of these unit vectors by another follows the cross-product rule. Thus,

$$\mathbf{ij} = \mathbf{k} \quad \mathbf{jk} = \mathbf{i} \quad \mathbf{ki} = \mathbf{j}$$

$$\tag{A.5}$$

$$\mathbf{ji} = -\mathbf{k} \quad \mathbf{kj} = -\mathbf{i} \quad \mathbf{ik} = -\mathbf{j}$$

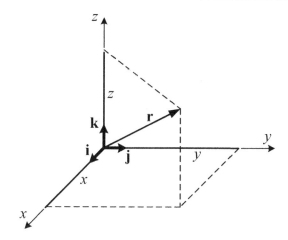

Fig. A.1 Unit vectors $\mathbf{i}, \mathbf{j}, \mathbf{k}$ are non-commutative

Using the results of Eqs. (A.5) and (A.4), we see that

$$\mathbf{ijk} = \mathbf{kk} = -1 \tag{A.6}$$

Combining Eqs. (A.4) and (A.6), we can write

$$\mathbf{ii} = \mathbf{jj} = \mathbf{kk} = \mathbf{ijk} = -1 \tag{A.7}$$

This result isn't obvious and it took Hamilton several years to recognize it. Recall that Hamilton was trying to extend complex numbers to higher dimensions, and on October 16, 1843, while walking to a meeting of the Royal Irish Academy, he had a flash of insight in which Eq. (A.7) popped into his head. He was so excited that as he passed the Brougham Bridge in Dublin, he took a knife and cut the formula in Eq. (A.7) into the stone. A plaque commemorating this event exists today on the bridge.

A *real* quaternion, t, is a quaternion in which the vector part is zero. Thus,

$$t = (s, 0) \tag{A.8}$$

A *pure* quaternion, p, is a quaternion in which the scalar part is zero. Thus,

$$p = (0, \mathbf{r})$$

(A.9)

A.2 Addition and Subtraction of Quaternions

Consider two quaternions, q_a and q_b, given by

$$q_a = s_a + \mathbf{r}_a = s_a + x_a \mathbf{i} + y_a \mathbf{j} + z_a \mathbf{k}$$

(A.10)

and

$$q_b = s_b + \mathbf{r}_b = s_b + x_b \mathbf{i} + y_b \mathbf{j} + z_b \mathbf{k}$$

(A.11)

The sum of the two quaternions q_a and q_b is given by

$$q_a + q_b = (s_a + \mathbf{r}_a) + (s_b + \mathbf{r}_b) = (s_a + s_b) + (\mathbf{r}_a + \mathbf{r}_b)$$

$$= (s_a + s_b) + (x_a + x_b)\mathbf{i} + (y_a + y_b)\mathbf{j} + (z_a + z_b)\mathbf{k}$$

(A.12)

The difference of the two quaternions q_a and q_b is given by

$$q_a - q_b = (s_a + \mathbf{r}_a) - (s_b + \mathbf{r}_b) = (s_a - s_b) + (\mathbf{r}_a - \mathbf{r}_b)$$

$$= (s_a - s_b) + (x_a - x_b)\mathbf{i} + (y_a - y_b)\mathbf{j} + (z_a - z_b)\mathbf{k}$$

(A.13)

A.3 Multiplication of Quaternions

If we multiply the two quaternions, q_a and q_b, given in Eqs. (A.10) and (A.11), and use the results of Eqs. (A.4) and (A.5), we can write

$$q_a q_b = (s_a + x_a \mathbf{i} + y_a \mathbf{j} + z_a \mathbf{k})(s_b + x_b \mathbf{i} + y_b \mathbf{j} + z_b \mathbf{k})$$

$$= (s_a s_b - x_a x_b - y_a y_b - z_a z_b)$$

$$+ (s_a x_b + s_b x_a + y_a z_b - z_a y_b)\mathbf{i}$$

$$+ (s_a y_b + s_b y_a + z_a x_b - x_a z_b)\mathbf{j}$$

$$+ (s_a z_b + s_b z_a + x_a y_b - y_a x_b)\mathbf{k}$$

(A.14)

which, using Eq. (A.2), can be reduced to

$$
\begin{aligned}
q_a q_b &= \left(s_a + \mathbf{r}_a \right)\left(s_b + \mathbf{r}_b \right) \\
&= s_a s_b - \mathbf{r}_a \cdot \mathbf{r}_b \\
&\quad + s_a \mathbf{r}_b + s_b \mathbf{r}_a + \mathbf{r}_a \times \mathbf{r}_b
\end{aligned}
\tag{A.15}
$$

where we have recalled from Eq. (3.3) that the cross product is given by

$$
\begin{aligned}
\mathbf{r}_a \times \mathbf{r}_b &= \begin{vmatrix} i & j & k \\ x_a & y_a & z_a \\ x_b & y_b & z_b \end{vmatrix} \\
&= \mathbf{i}\left(y_a z_b - z_a y_b \right) \\
&\quad + \mathbf{j}\left(z_a x_b - x_a z_b \right) \\
&\quad + \mathbf{k}\left(x_a y_b - y_a x_b \right)
\end{aligned}
\tag{A.16}
$$

Note from Eq. (A.15) that the product of two quaternions is itself a quaternion with the scalar part $\left(s_a s_b - \mathbf{r}_a \cdot \mathbf{r}_b \right)$ involving the dot product, and with a vector part $\left(s_a \mathbf{r}_b + s_b \mathbf{r}_a + \mathbf{r}_a \times \mathbf{r}_b \right)$ involving the cross product.

A.4 Conjugate, Norm, and Inverse of a Quaternion

The conjugate of the quaternion $q = s + \mathbf{r}$ is defined as

$$
q^* = s - \mathbf{r}
\tag{A.17}
$$

where the $+$ sign is replaced with a $-$ sign.

The *norm*, or magnitude squared, of the quaternion, q, is defined as

$$
|q|^2 = qq^*
\tag{A.18}
$$

Using Eqs. (A.17) and (A.15), we can write

$$qq^* = (s+\mathbf{r})(s-\mathbf{r})$$
$$= s^2 - \mathbf{r}\cdot(-\mathbf{r}) - s\mathbf{r} + s\mathbf{r} + \mathbf{r}\times\mathbf{r}$$
$$= s^2 + \mathbf{r}\cdot\mathbf{r}$$
$$= s^2 + x^2 + y^2 + z^2 \tag{A.19}$$

Thus,

$$|q|^2 = s^2 + x^2 + y^2 + z^2 \tag{A.20}$$

The *inverse* of the quaternion q is defined as

$$q^{-1} = \frac{q^*}{|q|^2} \tag{A.21}$$

Using Eq. (A.18), note that

$$qq^{-1} = \frac{qq^*}{|q|^2} = 1 \tag{A.22}$$

Dividing by a quaternion is done by multiplying by its inverse.

Consider a *unit* quaternion for which $|q|^2 = 1$. From Eq. (A.19), this means that

$$s^2 + \mathbf{r}\cdot\mathbf{r} - 1 \tag{A.23}$$

If $\mathbf{r} = w\mathbf{u}$, where $\mathbf{u}$ is a unit vector, then

$$\mathbf{r}\cdot\mathbf{r} = w^2 \tag{A.24}$$

Substituting Eq. (A.24) into Eq. (A.23) gives

$$s^2 + w^2 = 1 \tag{A.25}$$

We can let $s = \cos\theta$ and $w = \sin\theta$ in Eq. (A.25) because

$$\cos^2\theta + \sin^2\theta = 1 \tag{A.26}$$

Therefore, any unit quaternion can be written as

$$q = \cos\theta + \sin\theta\mathbf{u} \qquad (A.27)$$

Recall from Eq. (12.6) that Euler's equation is given by

$$e^{i\varphi} = \cos\varphi + i\sin\varphi \qquad (A.28)$$

Comparing Eqs. (A.27) and (A.28), by analogy, we can write a unit quaternion in exponential form as

$$q = e^{\mathbf{u}\theta} \qquad (A.29)$$

Note from Eq. (A.22) that if q is a unit vector for which $|q|^2 = 1$, then

$$q^{-1} = q* \qquad (A.30)$$

and the inverse of q is given by its conjugate.

Conjugate of a Product

Using Eqs. (A.15) and (A.17), note that we can write

$$\left(q_a q_b\right)^* = \left(s_a s_b - \mathbf{r}_a \cdot \mathbf{r}_b\right) - \left(s_a \mathbf{r}_b + s_b \mathbf{r}_a + \mathbf{r}_a \times \mathbf{r}_b\right) \qquad (A.31)$$

Now form the product

$$
\begin{aligned}
q_b^* q_a^* &= \left(s_b - \mathbf{r}_b\right)\left(s_a - \mathbf{r}_a\right) \\
&= \left(s_b s_a - \mathbf{r}_b \cdot \mathbf{r}_a\right) + \left[-s_b \mathbf{r}_a - s_a \mathbf{r}_b + \left(-\mathbf{r}_b\right)\times\left(-\mathbf{r}_a\right)\right] \\
&= \left(s_a s_b - \mathbf{r}_a \cdot \mathbf{r}_b\right) - \left(s_a \mathbf{r}_b + s_b \mathbf{r}_a + \mathbf{r}_a \times \mathbf{r}_b\right) \qquad (A.32)
\end{aligned}
$$

Comparing Eqs. (A.31) and (A.32), we see that

$$\left(q_a q_b\right)^* = q_b^* q_a^* \qquad (A.33)$$

Thus, the conjugate of a product is the product of the conjugates with the order of multiplication reversed.

A.5 Differentiation of a Quaternion

The "del" operator given in Eq. (6.22) can be written as

$$\nabla = \mathbf{i}\frac{\partial}{\partial x} + \mathbf{j}\frac{\partial}{\partial y} + \mathbf{k}\frac{\partial}{\partial z} \qquad (A.34)$$

Note that this is a vector, which we will take as the vector part of a pure quaternion **D** written as

$$\mathbf{D} = (0, \nabla) \qquad (A.35)$$

If we "multiply" Eq. (A.35) by the quaternion $q = (s, \mathbf{r})$ and use the quaternion multiplication rules in Eq. (A.15), we can write

$$\mathbf{D}q = (0, \nabla)(s, \mathbf{r})$$

or

$$\mathbf{D}q = -\nabla \cdot \mathbf{r} + \nabla s + \nabla \times \mathbf{r} \qquad (A.36)$$

Note that Eq. (A.36) is equivalent to

$$\mathbf{D}q = -div \ \mathbf{r} + \mathbf{grad} \ s + \text{curl} \ \mathbf{r} \qquad (A.37)$$

Thus, the differentiation of the quaternion $q = (s, \mathbf{r})$ results in a quaternion in which the scalar part is the divergence of $\mathbf{r}$ and the vector part is the sum of the gradient of s and the curl of $\mathbf{r}$.

A.6 Complex Quaternions

A real quaternion $q_r = s_a + \mathbf{r}_a$ is the sum of a real scalar plus a 3-dimensional real vector. These are the kinds of quaternions we have been considering up to this point. An imaginary quaternion $q_i = \iota(s_b + \mathbf{r}_b) = \iota s_b + \iota \mathbf{r}_b$ is a real quaternion multiplied by ι = iota, which is equal to $\sqrt{-1}$ and is the ordinary commutative imaginary unit of 2-dimensional complex numbers.

A complex quaternion q_c is the sum of a real quaternion and an imaginary quaternion. Thus, we can write

$$q_c = q_r + q_i = \left(s_a + \mathbf{r}_a\right) + \iota\left(s_b + \mathbf{r}_b\right) \tag{A.38}$$

or

$$q_c = \left(s_a + \iota s_b\right) + \left(\mathbf{r}_a + \iota \mathbf{r}_b\right) \tag{A.39}$$

A complex quaternion occupies an 8-dimensional space with units 1, ι, $\mathbf{i}$, $\mathbf{j}$, $\mathbf{k}$, $\iota\mathbf{i}$, $\iota\mathbf{j}$, $\iota\mathbf{k}$ along the real scalar, imaginary scalar, three real vector, and three imaginary vector axes respectively.

A.7 Matrix Representation of Quaternions

It is possible to represent the quaternion unit vectors $\mathbf{i}$, $\mathbf{j}$ and $\mathbf{k}$ defined by Eqs. (A.4) and (A.5) as 2×2 matrices with complex elements. These matrices are not unique. One possible representation is the following:

$$\mathbf{i} = \begin{bmatrix} \iota & 0 \\ 0 & -\iota \end{bmatrix} \qquad \mathbf{j} = \begin{bmatrix} 0 & 1 \\ -1 & 0 \end{bmatrix} \qquad \mathbf{k} = \begin{bmatrix} 0 & \iota \\ \iota & 0 \end{bmatrix} \tag{A.40}$$

In Eqs. (A.40), $\iota = \sqrt{-1}$ is the ordinary commutative imaginary unit of 2-dimensional complex numbers. Note that ordinary matrix multiplication will verify Eqs. (A.4) and (A.5). For example,

$$\mathbf{ij} = \begin{bmatrix} \iota & 0 \\ 0 & -\iota \end{bmatrix}\begin{bmatrix} 0 & 1 \\ -1 & 0 \end{bmatrix} = \begin{bmatrix} 0 & \iota \\ \iota & 0 \end{bmatrix} = \mathbf{k} \tag{A.41}$$

and

$$\mathbf{ii} = \begin{bmatrix} \iota & 0 \\ 0 & -\iota \end{bmatrix}\begin{bmatrix} \iota & 0 \\ 0 & -\iota \end{bmatrix} = \begin{bmatrix} -1 & 0 \\ 0 & -1 \end{bmatrix} = -1 \tag{A.42}$$

Using Eq. (A.40) and

$$1 = \begin{bmatrix} 1 & 0 \\ 0 & 1 \end{bmatrix} \tag{A.43}$$

we can write the quaternion from Eq. (A.3) as

$$q = s + x\mathbf{i} + y\mathbf{j} + z\mathbf{k}$$

$$= s \begin{bmatrix} 1 & 0 \\ 0 & 1 \end{bmatrix} + x \begin{bmatrix} \iota & 0 \\ 0 & -\iota \end{bmatrix} + y \begin{bmatrix} 0 & 1 \\ -1 & 0 \end{bmatrix} + z \begin{bmatrix} 0 & \iota \\ \iota & 0 \end{bmatrix}$$

$$= \begin{bmatrix} s + \iota x & y + \iota z \\ -y + \iota z & s - \iota x \end{bmatrix} \tag{A.44}$$

It is also possible to represent the quaternion unit vectors $\mathbf{i}, \mathbf{j}$ and $\mathbf{k}$ defined by Eqs. (A.4) and (A.5) as 4×4 matrices with real elements. These matrices are also not unique. One possible representation is the following:

$$1 = \begin{bmatrix} 1 & 0 & 0 & 0 \\ 0 & 1 & 0 & 0 \\ 0 & 0 & 1 & 0 \\ 0 & 0 & 0 & 1 \end{bmatrix} \qquad \mathbf{i} = \begin{bmatrix} 0 & 1 & 0 & 0 \\ -1 & 0 & 0 & 0 \\ 0 & 0 & 0 & 1 \\ 0 & 0 & -1 & 0 \end{bmatrix}$$

$$\mathbf{j} = \begin{bmatrix} 0 & 0 & -1 & 0 \\ 0 & 0 & 0 & 1 \\ 1 & 0 & 0 & 0 \\ 0 & -1 & 0 & 0 \end{bmatrix} \qquad \mathbf{k} = \begin{bmatrix} 0 & 0 & 0 & 1 \\ 0 & 0 & 1 & 0 \\ 0 & -1 & 0 & 0 \\ -1 & 0 & 0 & 0 \end{bmatrix} \tag{A.45}$$

Using matrix multiplication you can easily verify that Eqs. (A.4) and (A.5) are satisfied using the 4×4 matrices in Eqs. (A.45).

Using Eqs. (A.45) in Eq. (A.3), we can write a quaternion as the 4×4 matrix

$$q = s + x\mathbf{i} + y\mathbf{j} + z\mathbf{k} = \begin{bmatrix} s & x & -y & z \\ -x & s & z & y \\ y & -z & s & x \\ -z & -y & -x & s \end{bmatrix} \tag{A.46}$$

Appendix B

Fourier Transform of a Gaussian

B.1 Gaussian Function

Consider the Gaussian function given by

$$g(t) = e^{-\pi t^2} \tag{B.1}$$

A plot of this function is shown in Fig. B.1.

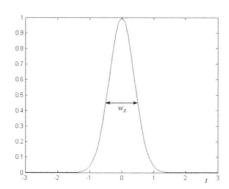

Fig. B.1 The Gaussian function, $g(t)$

The value of this function at $t = 1$ is 0.0432 and decreases rapidly thereafter. The area under this Gaussian function is given by the integral

$$I = \int_{-\infty}^{\infty} e^{-\pi t^2} dt \tag{B.2}$$

One way to evaluate the integral in Eq. (B.2) is to square it.

$$I^2 = \int\limits_{-\infty}^{\infty} e^{-\pi x^2} dx \int\limits_{-\infty}^{\infty} e^{-\pi y^2} dy = \int\limits_{-\infty}^{\infty} \int\limits_{-\infty}^{\infty} e^{-\pi(x^2+y^2)} dxdy \qquad (B.3)$$

Now convert to polar coordinates by letting

$$x = r\cos\theta \qquad y = r\sin\theta \qquad (B.4)$$

Then, Eq. (B.3) can be written as

$$I^2 = \int\limits_{0}^{\infty} \int\limits_{0}^{2\pi} e^{-\pi r^2} r d\theta dr$$

$$= 2\pi \int\limits_{0}^{\infty} re^{-\pi r^2} dr \qquad (B.5)$$

To integrate Eq. (B.5), let

$$\alpha = r^2 \qquad d\alpha = 2rdr \qquad (B.6)$$

and rewrite Eq. (B.5) as

$$I^2 = 2\pi \int\limits_{0}^{\infty} e^{-\pi\alpha} d\alpha = 2\pi \left(-\frac{1}{2\pi} \right) e^{-\pi\alpha} \Big|_{0}^{\infty} = 1 \qquad (B.7)$$

Therefore, the area under the Gaussian function, given by Eq. (B.2) is

$$Area = \int\limits_{-\infty}^{\infty} e^{-\pi t^2} dt = 1 \qquad (B.8)$$

We will take the width of the Gaussian function, w_g, in Fig. B.1 to be 1, between the values of $t = -1/2$ and $t = +1/2$. This will occur at a value of the Gaussian function equal to $e^{-\pi/4} = 0.4559$. A Gaussian function of width $w_g = a$ will be given by

$$g(t/a) = \frac{1}{a} e^{-\pi(t/a)^2} \tag{B.9}$$

An alternate, common definition of the Gaussian function is

$$p(t) = \frac{1}{\sqrt{2\pi\sigma^2}} e^{-t^2/2\sigma^2} \tag{B.10}$$

where σ^2 is the variance and σ is the standard deviation. Comparing Eqs. (B.9) and (B.10), the Gaussian width a in Eq. (B.9) is related to σ by the equation

$$a = \sigma\sqrt{2\pi} \tag{B.11}$$

As we will see in the next section, the definition of the Gaussian function given by Eq. (B.1) has the advantage of producing the same function for its Fourier transform.

B.2 Fourier Transform of a Gaussian Function

The Fourier transform, $G(f)$, of a function $g(t)$ is defined as

$$G(f) = \int_{-\infty}^{\infty} g(t) e^{-j2\pi ft} dt \tag{B.12}$$

and the inverse transform is

$$g(t) = \int_{-\infty}^{\infty} G(f) e^{j2\pi ft} df \tag{B.13}$$

Let $g(t)$ be the Gaussian function given by Eq. (B.1). Then, from Eq. (B.12), its Fourier transform will be

$$G(f) = \int_{-\infty}^{\infty} e^{-\pi t^2} e^{-j2\pi ft} dt \tag{B.14}$$

To find this integral, it is convenient to first differentiate Eq.

(B.14) with respect to f. Thus,

$$\frac{dG(f)}{df} = \int_{-\infty}^{\infty} e^{-\pi t^2} \left(-j2\pi t\right) e^{-j2\pi ft} dt \tag{B.15}$$

Proceed by integrating by parts. Let

$$u = e^{-j2\pi ft} \qquad du = \left(-2\pi jf\right) e^{-2\pi jft} dt \tag{B.16}$$

$$dv = -2\pi jt e^{-\pi t^2} dt \qquad v = je^{-\pi t^2} \tag{B.17}$$

Using Eqs. (B.16) and (B.17) in the integration by parts formula

$$\int_{-\infty}^{\infty} udv = uv\Big|_{-\infty}^{\infty} - \int_{-\infty}^{\infty} vdu \tag{B.18}$$

Eq. (B.15) becomes

$$\frac{dG(f)}{df} = -\int_{-\infty}^{\infty} je^{-\pi t^2} \left(-j2\pi f\right) e^{-2\pi jft} dt$$

$$= -2\pi f \int_{-\infty}^{\infty} e^{-\pi t^2} e^{-j2\pi ft} dt$$

which, using Eq. (B.14), can be written as

$$\frac{dG(f)}{df} = -2\pi fG(f) \tag{B.19}$$

The solution of Eq. (B.19) is

$$G(f) = G(0)e^{-\pi f^2} \tag{B.20}$$

where

$$G(0) = \int_{-\infty}^{\infty} g(t)dt = \int_{-\infty}^{\infty} e^{-\pi t^2} dt = 1 \tag{B.21}$$

where Eq. (B.8) was used in the last step. Therefore, the Fourier transform of the Gaussian function in Eq. (B.1) is

$$G(f) = e^{-\pi f^2} \tag{B.22}$$

We will use the symbol $\Leftrightarrow$ to denote a Fourier transform pair. Thus,

$$g(t) \Leftrightarrow G(f) \tag{B.23}$$

is read as, " $g(t)$ has as its Fourier transform $G(f)$, and $G(f)$ has as its inverse Fourier transform $g(t)$."

Therefore, for a Gaussian function,

$$e^{-\pi t^2} \Leftrightarrow e^{-\pi f^2} \tag{B.24}$$

The Fourier transform scaling theorem states that

$$g(t/a) \Leftrightarrow |a|G(af) \tag{B.25}$$

and the shifting theorem states that

$$g(t - t_0) \Leftrightarrow G(f)e^{-j2\pi f t_0} \tag{B.26}$$

We can therefore write the Fourier transform of a Gaussian of width a as

$$e^{-\pi(t^2/a^2)} \Leftrightarrow |a|e^{-\pi a^2 f^2} \tag{B.27}$$

If this scaled Gaussian is shifted by t_0, then its Fourier transform is given by

$$e^{-\pi \frac{(t-t_0)^2}{a^2}} \Leftrightarrow |a|e^{-\pi a^2 f^2}e^{-j2\pi f t_0} \tag{B.28}$$

Index

Made in the USA
Middletown, DE
26 August 2020